# OpenStack
## 从零开始学

卢万龙 周萌 编著

电子工业出版社
Publishing House of Electronics Industry
北京·BEIJING

## 内容简介

OpenStack 作为开源云计算技术首当其冲，有着广泛的受众、活跃的社区和良好的传播，尊为云计算技术的领导者。

本书由浅入深，从设计理论到实际操作，带领读者认识 OpenStack 云计算的全貌，轻松步入 OpenStack 云计算的世界。其内容涵盖了 OpenStack 云计算设计理论，虚拟化技术 KVM 和 Xen 的原理与应用，4 种 OpenStack 网络架构（Flat、Local、GRE 和 VXLAN）模式和网络 OSI 7 层模型介绍，Ceph 分布式存储，OpenStack 安装配置（Nova、Cinder、Neutron、Horizon、Swift 和 Keystone 等服务组件）、应用场景和实际操作（卷管理、创建网络和实例、实例热迁移和冷迁移）等多个方面，使读者读后如沐春风，真正喜欢云计算这项技术。

本书适合刚刚或者计划进入云计算领域的初级读者学习，也适合已经进入云计算领域并且有一定相关知识或认识的中级读者阅读。对于一些从事售前工作的读者，本书也非常适用。

未经许可，不得以任何方式复制或抄袭本书之部分或全部内容。
版权所有，侵权必究。

**图书在版编目（CIP）数据**

OpenStack 从零开始学 / 卢万龙，周萌编著. —北京：电子工业出版社，2016.11
ISBN 978-7-121-29873-8

Ⅰ．①O… Ⅱ．①卢… ②周… Ⅲ．①计算机网络 Ⅳ．①TP393

中国版本图书馆 CIP 数据核字（2016）第 215705 号

策划编辑：孙学瑛
责任编辑：徐津平
特约编辑：赵树刚
印　　刷：北京盛通商印快线网络科技有限公司
装　　订：北京盛通商印快线网络科技有限公司
出版发行：电子工业出版社
　　　　　北京市海淀区万寿路 173 信箱　　邮编：100036
开　　本：787×980　1/16　　印张：22.25　　字数：428 千字
版　　次：2016 年 11 月第 1 版
印　　次：2023 年 7 月第 16 次印刷
定　　价：69.00 元

凡所购买电子工业出版社图书有缺损问题，请向购买书店调换。若书店售缺，请与本社发行部联系，联系及邮购电话：（010）88254888，88258888。
质量投诉请发邮件至 zlts@phei.com.cn，盗版侵权举报请发邮件至 dbqq@phei.com.cn。
本书咨询联系方式：010-51260888-819，faq@phei.com.cn。

# 序一

第一次接到学生的作序邀请,有些许惊讶,惊讶之余,多有感慨!感慨有三:一是当初混沌懵懂的大学生已经在社会上茁壮成长,已不复当年记忆;二是自己的学生有如此进步,实为其感到高兴;三是自己的学生还在从事本专业工作,坚持不懈,为自己多年的教学生涯感到慰藉。遥记当年和学生第一次通过邮件讨论 Java 问题,后期做毕业设计,研究项目和语言架构,学生均付之于全力。以观如今,也属自然,望之继续努力。

当今,云计算已开始浸入和影响我们的生活。社会大众所使用的微博、微信、支付宝等手机 APP 多半在云端运行,大多数人的日常生活都已离不开它们,云计算俨然成为我们生活和学习的必需品。正如工业革命时期的水、电和交通一样,云计算正在让信息技术和信息服务实现社会化、集约化和专业化,让信息服务成为社会的公共基础设施。作为一种新型的服务提供模型,云计算让用户可随时、随地、按需地通过网络访问计算、存储等各种共享资源。

短短十余载,云计算从概念实现了落地,从星星之火演变成燎原烈火。在云计算的 IaaS 领域,OpenStack 风头正盛,社区活跃、灵活性、创新性、不被厂家锁定和良好的生态环境等一系列优点促使它成为云计算的"宠儿"。在云计算的发展进程中,涌现出众多 OpenStack 的产品、方案和服务。各大、中、小型公司竞相趋之,希望能够借助 OpenStack 这个风口而起飞。

苏格拉底有千年一问,研究哲学讲究三要素,问和研都集中在这三个问题上:我是谁?我从哪里来?我要到哪里去?本书作者抱着回答这三个问题的态度,并在此落笔,从理论出发，结合实际，综合实操，全面讲解 OpenStack，阐述了 OpenStack 的前世今生，详述了 OpenStack 的何去何从。希望广大读者能够和本书作者一起，带着这三个问题阅读此书，理解云计算和学习 OpenStack，重温快乐学习的过程。谢谢！

谢斌红

太原科技大学计算机科学与技术学院副教授

# 序二

与本书作者的初次相识,是在一次云计算技术交流大会上,我俩一见如故,相谈甚欢。本书作者对技术的追求和痴迷,给我留下了深刻的印象,也一下子拉近了我们的距离。同为技术研究人员的我们,此后一直保持联系,对技术的钻研、实践中问题的探讨、对行业发展的思考……最近一次见面,本书作者拿着自己的书稿,让我帮忙写一段话。我在浏览此书后,觉得此书对于云计算的初级学者和中级读者来讲,在系统、全面地学习OpenStack方面是非常有帮助的。

OpenStack作为一款云计算软件,成功地利用软件定义了传统的一切,包括软件定义存储、软件定义网络等。它将传统所见的黑盒子全部利用软件定义成白盒子。原来的计算或存储设备很神秘,像"变形金刚"一样屹立在我们的机房中,默默地提供服务,而数据在这些设备中如何进行计算、存储,以及数据的安全性、持久性,只能依赖厂商的产品说明书,而我们被拒之门外。然而OpenStack解决了这一切,只要你懂得代码,你就通晓一切。OpenStack对计算、存储、网络等多个功能进行了融合,成为一个完整的个体,解决了我们的各种IT需求。在原来的世界中,计算是计算、存储是存储、网络是网络,彼此之间分离,安装、配置、部署和运维极其复杂,并且还要考虑不同设备之间的兼容性。OpenStack将这些功能设计成不同的服务组件,彼此之间松耦合,组合在一起,并且提供了一套完整的管理方法和工具,完成了我们想要实现的所有功能,承载它们的物理硬件也由"变形金刚"变成了普通的"铁盒子"——X86服务器。开源是OpenStack最大的优势,避免了被IT厂商锁定,我们可以将核心技术掌握在自己手中,再也不用因为某个技术不熟悉而弃用或另行招聘人才,或者因为不兼容而不得不购买固定品牌的昂贵设备,避免由此造成企业成本的增加。

回想自己初入云计算领域，当时亦是十分茫然，在市面上找了很多相关书籍，如饥似渴地阅读。本书作为一本系统化讲述 OpenStack 的书籍，阐述了 OpenStack 的前世今生，详述了 OpenStack 的何去何从，深入浅出，理论结合实践，具有很强的实操性。相信初入云计算领域的技术人员在详细阅读此书后，会有一种打通任督二脉、茅塞顿开的收获；对于中级读者来说也能融会贯通，有新的收获。祝愿读者在此书中有一个愉快的阅读旅程！

张志飞

WatchGuard 加拿大区首席架构师

# 前　言

记得初识云计算时，根本不知道云计算为何物，同时出现的还有物联网概念，它们满满地冲击着我这个刚刚适应工作岗位的菜鸟。随着各IT厂商对云计算概念的热炒，慢慢地我知道了什么是云计算、云计算能够做什么。不知道是我抽离现象看到了本质，还是目光浅薄，竟然逐渐看轻云计算，认为其不过是IT厂商忽悠甲方的一个概念，要落地并产生效益很难。

后来，中国整个IT市场掀起了去IOE的飓风，尤其是在"棱镜门"爆出后，呈愈演愈烈之势，传统的高性能、高稳定性和高安全性设备与架构失去了其继续存在的依靠，让传统IT建设在整个IT大环境中显得那么格格不入。同时，也让我这个传统IT从业者变得惴惴不安，让我多年视若瑰宝的UNIX和存储技术一下子变成了过去时。在我情绪低落时，无所适从的感觉油然而生，这种感觉既迷茫又无奈。

这也印证了一句名言：世界上唯一不变的就是变化。人们大抵只会在互相调笑的时候才会祭出"三观不正确"的观点来反击他人，但是在某段比较盲从的时间里，可能也只有一个人具备的先天与后天共同作用而产生的"三观"能帮助你、解救你，至少自己可以这么认为。正确的"三观"告诉我们，人要顺应潮流，要顺势而为，要迎合变化，要拥抱改变。因此，我放弃了学习多年的UNIX和存储技术，投身于Linux和各种开源技术，也正在此时，OpenStack进入了我的世界。我重拾荒废多年的Linux，开始学习Hadoop和OpenStack。

接触OpenStack三年多来，我一直在碎片化地学习，走了不少弯路，也错误地理解了很多概念。人多半都是在困惑的道路上一直前行，走走看看，看看走走。某天下午，我无意中看到一句话——最好的学习就是写作，如梦初醒，于是萌生了写作此书的想法。希望阅读此书的读者能够在学习 OpenStack 的道路上少走一些弯路，能够系统、全面地学习和掌握 OpenStack。

## 本书内容

本书共分为 3 篇 28 章。第 1 篇是原理篇，包括 9 个章节，详细讲述了 OpenStack 项目中各服务组件的原理知识，对计算虚拟化技术、网络知识模型和分布式存储进行了重点讲解，对读者学习、理解 OpenStack 很有帮助，同时为下一篇的学习做好铺垫。

第 2 篇是安装配置篇，包括 11 个章节，详细讲述了如何安装配置 OpenStack 项目中的各服务组件。该篇承上启下，既是对原理篇所述原理知识的落地，又是对管理篇操作由来的一个剖析。

第 3 篇是管理篇，包括 8 个章节，详细讲述了如何管理 OpenStack 环境，既包括管理主机类型、虚拟机实例这样的基础操作，又包括虚拟机实例热迁移、冷迁移、转移及 OpenStack 版本升级这样的高级操作，相信对读者完全理解和掌握 OpenStack 会有一个极大的促进。

## 读者对象

本书适合刚刚或者计划进入云计算领域的初级读者学习，也适合已经进入云计算领域并且有一定相关知识或认识的中级读者阅读。对于一些从事售前工作的读者，本书也非常适用。同时，非常欢迎一些云计算资深人士或行业专家阅读本书以提出宝贵意见。

## 感谢

感谢张鹏程，鹏程在工作方式方法、为人处世方面给了我很多影响，对我未来的职业生涯有极大的帮助。

感谢马筵峰，老马给予了我极大的鼓励和支持，使我始终存有对技术的兴趣，坚持前行。

感谢瑞飞的领导和兄弟们，包括杨剑锋书记、陆春阳副总经理、罗革新副总经理、刘哲生总监、丁闯总监、丁建新、张骁、曾国强、张树海、杨友红、侯明扬、金鹏飞、王志刚，还有一些没有提到的兄弟姐妹，你们在工作和生活上的帮助和指导让我受益良多，非常感谢！

感谢孙学瑛老师，孙老师认真的工作态度、专业的出版知识和热心的帮助，使得此书最终可以出版。

感谢周琦老师，周老师热心的帮助和专业的项目管理知识，对本书的写作和出版给予了巨大帮助。

## 声明

1. 本书部分图片来自互联网，版权归原作者所有，引用的目的是使读者更好地理解此书内容，感谢原作者贡献。

2. 作者水平有限，开源技术无限，如有错误，敬请广大读者斧正。

## 感悟

1. IT 技术变化之快，摧枯拉朽。技术人得闲时，还是要多看看、多想想、多听听。

2. 许上等愿，结中等缘，享下等福。与诸君共勉。

# 目 录

## 第 1 篇 原理篇

**第 1 章 云计算之 OpenStack** .................................................. 2
    1.1 什么是云 .................................................................... 2
    1.2 OpenStack 发展历程 ........................................................ 4
    1.3 OpenStack 概念设计 ........................................................ 4
    1.4 OpenStack 服务组件 ........................................................ 6

**第 2 章 计算（Nova）服务介绍** ................................................ 8
    2.1 架构设计 .................................................................. 10
    2.2 虚拟化技术介绍 ............................................................ 10
        2.2.1 KVM 虚拟化技术 .................................................. 12
        2.2.2 Xen 虚拟化技术 ................................................... 22
    2.3 Libvirt 技术介绍 ........................................................... 31
        2.3.1 Libvirt API 介绍 .................................................. 32
        2.3.2 Libvirt 网络架构 .................................................. 34
        2.3.3 Libvirt 存储架构 .................................................. 36

**第 3 章 网络（Neutron）服务介绍** ............................................ 37
    3.1 网络 OSI 7 层模型 .......................................................... 38
    3.2 网络介绍 .................................................................. 43

| | | |
|---|---|---|
| 3.3 | 网络架构 | 44 |
| 3.4 | 网络 API 简介 | 45 |
| 3.5 | LBaaS 和 FWaaS | 46 |
| 3.6 | 网络类型介绍 | 48 |

## 第 4 章 存储服务介绍 ..... 51

| | | |
|---|---|---|
| 4.1 | 块存储（Cinder）服务介绍 | 53 |
| 4.2 | 对象存储（Swift）服务介绍 | 54 |
| | 4.2.1 对象存储特点 | 55 |
| | 4.2.2 对象存储组成 | 57 |
| 4.3 | 文件系统存储 | 62 |
| 4.4 | Ceph 简介 | 62 |
| | 4.4.1 存储数据过程 | 64 |
| | 4.4.2 可扩展性和高可用性 | 65 |
| | 4.4.3 集群管理 | 68 |

## 第 5 章 计量（Ceilometer）服务介绍 ..... 72

| | | |
|---|---|---|
| 5.1 | 计量服务组件组成 | 72 |
| 5.2 | 计量服务组件支持列表 | 73 |

## 第 6 章 身份认证（Keystone）服务介绍 ..... 75

## 第 7 章 镜像（Glance）服务介绍 ..... 78

## 第 8 章 仪表板（Horizon）服务介绍 ..... 80

## 第 9 章 编排（Heat）服务介绍 ..... 87

# 第 2 篇　安装配置篇

## 第 10 章 OpenStack 安装配置准备 ..... 90

| | | |
|---|---|---|
| 10.1 | 架构设计 | 90 |
| 10.2 | 基础环境准备 | 93 |

|     |        |                                        |     |
| --- | ------ | -------------------------------------- | --- |
|     | 10.2.1 | 安全设置规则                           | 95  |
|     | 10.2.2 | 主机节点网络设置                       | 96  |
|     | 10.2.3 | 节点时钟同步                           | 102 |
|     | 10.2.4 | 配置 OpenStack 安装源和运行环境        | 104 |
|     | 10.2.5 | 安装和配置 SQL 数据库                  | 104 |
|     | 10.2.6 | 安装和配置 NoSQL 数据库                | 105 |
|     | 10.2.7 | 安装和配置消息队列                     | 106 |

## 第 11 章 身份认证（Keystone）服务安装配置 ... 108

| 11.1 | 安装和配置 | | 108 |
| --- | --- | --- | --- |
| 11.2 | 创建 service entity 和 API endpoint | | 112 |
|  | 11.2.1 | 准备 | 113 |
|  | 11.2.2 | 创建过程 | 113 |
| 11.3 | 创建项目、用户和角色 | | 115 |
| 11.4 | 检查配置 | | 118 |
| 11.5 | 定义 OpenStack 客户端环境变量脚本 | | 119 |
|  | 11.5.1 | 创建环境变量脚本 | 120 |
|  | 11.5.2 | 验证 | 120 |

## 第 12 章 镜像（Glance）服务安装配置 ... 122

| 12.1 | 安装和配置 | | 122 |
| --- | --- | --- | --- |
|  | 12.1.1 | 准备 | 122 |
|  | 12.1.2 | 安装和配置 Glance 镜像服务组件 | 125 |
|  | 12.1.3 | 安装完成 | 127 |
| 12.2 | 验证 | | 128 |

## 第 13 章 计算（Nova）服务安装配置 ... 130

| 13.1 | 安装和配置（控制节点） | | 130 |
| --- | --- | --- | --- |
|  | 13.1.1 | 准备 | 130 |
|  | 13.1.2 | 安装和配置 Nova 计算服务组件 | 133 |
|  | 13.1.3 | 安装完成 | 136 |

| | | |
|---|---|---|
| 13.2 | 安装和配置（计算节点） | 136 |
| | 13.2.1 安装和配置 Nova 计算服务组件 | 136 |
| | 13.2.2 安装完成 | 139 |
| 13.3 | 验证 | 139 |

## 第 14 章 网络（Neutron）服务安装配置 … 141

| | | |
|---|---|---|
| 14.1 | 安装和配置（控制节点） | 141 |
| | 14.1.1 准备 | 142 |
| | 14.1.2 配置 Neutron 网络服务组件 | 144 |
| | 14.1.3 配置 metadata agent | 156 |
| | 14.1.4 配置计算服务组件 | 157 |
| | 14.1.5 安装完成 | 157 |
| 14.2 | 安装和配置（计算节点） | 158 |
| | 14.2.1 网络服务组件安装和配置通用组件 | 158 |
| | 14.2.2 配置网络核心组件 | 160 |
| | 14.2.3 配置计算服务组件 | 162 |
| | 14.2.4 安装完成 | 163 |
| 14.3 | 验证 | 163 |

## 第 15 章 仪表板（Horizon）服务安装配置 … 165

| | | |
|---|---|---|
| 15.1 | 安装和配置 | 165 |
| | 15.1.1 安装和配置 Horizon 仪表板服务组件 | 166 |
| | 15.1.2 安装完成 | 168 |
| 15.2 | 验证 | 168 |

## 第 16 章 块存储（Cinder）服务安装配置 … 169

| | | |
|---|---|---|
| 16.1 | 安装和配置（控制节点） | 169 |
| | 16.1.1 准备 | 170 |
| | 16.1.2 安装和配置 Cinder 块存储服务组件 | 173 |
| | 16.1.3 安装完成 | 175 |

## 16.2 安装和配置（存储节点） .................................................. 176
### 16.2.1 准备 .................................................. 176
### 16.2.2 安装和配置 Cinder 块存储服务组件 .................................................. 177
### 16.2.3 安装完成 .................................................. 180
## 16.3 验证 .................................................. 180

# 第 17 章 对象存储（Swift）服务安装配置 .................................................. 181
## 17.1 安装和配置（控制节点） .................................................. 181
### 17.1.1 准备 .................................................. 182
### 17.1.2 安装和配置 Swift 对象存储服务组件 .................................................. 184
## 17.2 安装和配置（存储节点） .................................................. 186
### 17.2.1 准备 .................................................. 186
### 17.2.2 安装和配置 Swift 对象存储服务组件 .................................................. 188
## 17.3 创建和分发 Ring .................................................. 190
### 17.3.1 创建用户 Ring .................................................. 190
### 17.3.2 创建 Container Ring .................................................. 192
### 17.3.3 创建 Object Ring .................................................. 193
### 17.3.4 分发 Ring 配置文件 .................................................. 195
## 17.4 安装完成 .................................................. 195
## 17.5 验证 .................................................. 197

# 第 18 章 编排（Heat）服务安装配置 .................................................. 199
## 18.1 安装和配置 .................................................. 199
### 18.1.1 准备 .................................................. 199
### 18.1.2 安装和配置 Heat 编排服务组件 .................................................. 205
### 18.1.3 安装完成 .................................................. 208
## 18.2 验证 .................................................. 208

# 第 19 章 计量（Ceilometer）服务安装配置 .................................................. 209
## 19.1 安装和配置 .................................................. 209
### 19.1.1 准备 .................................................. 209

19.1.2 安装和配置 Ceilometer 计量服务组件 ............213
19.1.3 安装完成 ............215
19.2 启用 Glance 镜像服务计量 ............215
19.3 启用 Nova 计算服务计量 ............216
19.3.1 安装和配置 agent ............216
19.3.2 配置 Nova 计算服务使用 Ceilometer 计量服务 ............218
19.4 启用 Cinder 块存储服务计量 ............218
19.5 启用 Swift 对象存储服务计量 ............219
19.5.1 准备 ............219
19.5.2 配置 Swift 对象存储服务使用 Ceilometer 计量服务 ............220
19.6 验证 ............220

## 第 20 章 建立虚拟机实例测试 ............222

20.1 创建虚拟网络 ............222
20.1.1 架构一网络（Public Provider Network）............222
20.1.2 架构二网络（Private Project Network）............225
20.2 创建 Key Pair ............231
20.3 创建 Security Group 规则 ............232
20.4 创建虚拟机实例 ............232
20.4.1 创建虚拟机实例（Public Provider Network）............232
20.4.2 创建虚拟机实例（Private Project Network）............238
20.5 创建块存储 ............243

# 第 3 篇 管理篇

## 第 21 章 OpenStack 项目管理 ............250

21.1 管理租户、用户和角色 ............250
21.1.1 命令行方式 ............251
21.1.2 图形界面方式 ............255

## 21.2 管理主机类型 ... 257
### 21.2.1 命令行方式 ... 257
### 21.2.2 图形界面方式 ... 259
## 21.3 管理安全组 ... 263
## 21.4 管理主机集合 ... 265
## 21.5 资源使用率统计 ... 267
### 21.5.1 命令行方式 ... 267
### 21.5.2 图形界面方式 ... 269
## 21.6 查看系统服务信息 ... 270

# 第 22 章 仪表板使用 ... 273
## 22.1 Logo 和图形界面定制化 ... 273
## 22.2 HTML 标题、Logo 链接和帮助定制化 ... 275

# 第 23 章 管理镜像 ... 276
## 23.1 命令行方式 ... 276
## 23.2 图形界面方式 ... 279

# 第 24 章 管理网络 ... 282
## 24.1 命令行方式 ... 282
## 24.2 图形界面方式 ... 285

# 第 25 章 管理卷设备 ... 287
## 25.1 命令行方式 ... 287
## 25.2 图形界面方式 ... 295

# 第 26 章 管理虚拟机实例 ... 298
## 26.1 创建虚拟机实例 ... 298
### 26.1.1 命令行方式 ... 298
### 26.1.2 图形界面方式 ... 302
## 26.2 操作虚拟机实例 ... 303
### 26.2.1 命令行方式 ... 303
### 26.2.2 图形界面方式 ... 306

26.3 选择主机节点运行实例 ............................................................. 307
26.4 计算节点配置 SSH 互信 ........................................................... 308
26.5 实例热迁移 ............................................................................. 310
    26.5.1 KVM ........................................................................... 311
    26.5.2 XenServer ..................................................................... 317
26.6 实例冷迁移 ............................................................................. 318
26.7 实例转移 ................................................................................. 319

## 第 27 章 OpenStack 版本升级 .................................................. 322
27.1 升级准备 ................................................................................. 322
27.2 版本升级 ................................................................................. 325
27.3 版本回退 ................................................................................. 329

## 第 28 章 故障排查 ..................................................................... 333
28.1 计算服务组件故障排查 ............................................................. 333
28.2 块存储服务组件故障排查 ......................................................... 335

# 第 1 篇

# 原 理 篇

- 第 1 章 云计算之 OpenStack
- 第 2 章 计算（Nova）服务介绍
- 第 3 章 网络（Neutron）服务介绍
- 第 4 章 存储服务介绍
- 第 5 章 计量（Ceilometer）服务介绍
- 第 6 章 身份认证（Keystone）服务介绍
- 第 7 章 镜像（Glance）服务介绍
- 第 8 章 仪表板（Horizon）服务介绍
- 第 9 章 编排（Heat）服务介绍

# 第 1 章
# 云计算之 OpenStack

云计算是互联网发展进程中的一个阶段，提供了用户所需的一切，从计算能力、计算基础设施到应用、业务流程等，作为一种服务，在任意时间、任意地点交付使用。

从发展过程看，云计算已经完全改变了公司利用技术服务于客户、合作伙伴和供应商的方式，提供了更高效、更具成本优势的 IT 服务。目前云计算可以以三种不同的形式提供服务，分别为公有云、私有云和混合云。公有云是企业将自己的内部 IT 资源服务于外部客户，外部客户具有使用权、无所有权，主要提供商有 AWS、Azure 和阿里云等。私有云是企业将自己的 IT 资源服务于自己，专为自己服务而构建，具有使用权和所有权，主要技术包括 OpenStack 和 CloudStack 等。混合云，顾名思义，其融合了公有云和私有云，企业根据不同的业务需求而综合利用公有云和私有云提供的服务，解决方案提供商有 IBM、华为等。

## 1.1 什么是云

开篇之初，笔者认为有必要介绍一下什么是云，尽管现在云发展得如火如荼，并且逐渐成熟，在多个地区和公司已经落地，并且拥有许多应用场景。基于此，更需要大家对云有一个正确的认识。

什么是云？或者说云的核心特征是什么？有人说云具备超强的计算能力，对吗？笔者认为正确，大量 CPU 和内存进行池化，已然具备超乎寻常的计算能力。有人说云具备弹性扩展能力，对吗？笔者认为正确，你看专业人士谈云必提的无停机时间横向扩展、嗤之以鼻且缺点多多的纵向扩展。更有人说云分多种，有 IAAS，代表基础设施云；有 PAAS，代表平台云；有 SAAS，代表软件云；还有 CAAS，等等，对吗？笔者认为正确。有人说云代表着软件定义一切，对吗？笔者认为也正确，现在大家都在谈软件定义网络、软件定义存储等，由此也佐证了开发为王、软件定义的强大。那说了这么多，讲清楚云了吗？笔者认为还没有。

那到底如何理解云呢？笔者认为可以从三方面来理解。

第一，云的构成。

- 用户：用户不需要知道关于底层技术的任何事情，只需要知道利用云如何实现自己的业务诉求。
- 商业管理：对云中数据和服务的治理提供完善的商业管理规则，云提供商提供可预测和可保证的 SLA 和安全协议。
- 云提供商：云提供商对 IT 资产运行和维护负责。

第二，云的特点。

- 具备弹性扩展能力。
- 提供自助服务功能。
- 具备标准程序接口（Application Programming Interfaces，APIs）。
- 拥有付费和计量功能。

第三，云的定义。

云是一种服务，其本质就是为用户提供优质服务，利用虚拟化技术、分布式技术、软件定义等技术为用户提供便捷、简单、准确、可用、按需所取的服务。

## 1.2 OpenStack 发展历程

OpenStack 是在 2010 年由托管服务器及云计算提供商 RackSpace 和美国航天航空局 NASA 共同发起的开源项目，愿景是为数据中心建立一套云操作系统，大幅提升数据中心的运营效率。其旨在为公有云和私有云的建设与管理提供软件开源项目，社区拥有超过 170 家企业及 3000 位开发者，这些机构与个人都将 OpenStack 作为基础设施即服务（IaaS）的解决方案。下表是 OpenStack 已经发布的版本信息及使用状态。

| 版 本 | 状 态 | 生命周期开始日期 | 生命周期结束日期 |
| --- | --- | --- | --- |
| Ocata | | 未知 | 未知 |
| Newton | 正在开发 | 预计 2016-10-16 | 未知 |
| Mitaka | 当前最新、最稳定版本，提供支持 | 2016-04-07 | 未知 |
| Liberty | 提供支持 | 2015-10-15 | 2016-11-17 |
| Kilo | 生命周期已结束 | 2014-04-30 | 2016-05-02 |
| Juno | 生命周期已结束 | 2014-10-16 | 2015-12-07 |
| Icehouse | 生命周期已结束 | 2014-04-17 | 2015-07-02 |
| Havana | 生命周期已结束 | 2013-10-17 | 2014-09-30 |
| Grizzly | 生命周期已结束 | 2013-04-04 | 2014-03-29 |
| Folsom | 生命周期已结束 | 2012-09-27 | 2013-11-19 |
| Essex | 生命周期已结束 | 2012-04-05 | 2013-05-06 |
| Diablo | 生命周期已结束 | 2011-09-22 | 2013-05-06 |
| Cactus | 不支持 | 2011-04-15 | |
| Bexar | 不支持 | 2011-02-03 | |
| Austin | 不支持 | 2010-10-21 | |

## 1.3 OpenStack 概念设计

OpenStack 是一个开源的云计算平台，由来自全世界的开发者创造了 OpenStack 开源项目，社区活跃度和代码贡献量非常高，主要特点是易实施、大规模弹性扩展和功能丰富。下图展示了 OpenStack 项目中各服务组件之间的关系。

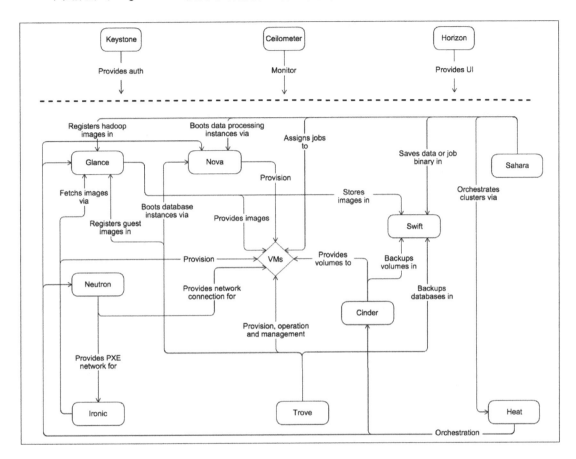

各服务组件的功能如下：

- Keystone 为各服务组件提供用户认证和权限验证功能。

- Ceilometer 为各服务组件提供监控、检索和计量功能。

- Horizon 为用户操作 OpenStack 项目中的各服务组件提供基于 Web 形式的图形界面。

- Glance 为虚拟机实例提供镜像服务，同时，Glance 服务中的镜像介质存放在 Swift 中。

- Neutron 为虚拟机实例提供网络连接服务，同时也为 Ironic 提供 PXE 网络。

- Ironic 提供物理机的添加、删除、电源管理和安装部署等功能。
- Nova 为虚拟机实例提供计算资源。
- Trove 为虚拟机镜像提供注册服务；使用 Nova 启动数据库实例；依附虚拟机实例，提供数据存储、操作和管理；可以备份数据库实例到 Swift 中。
- Cinder 为虚拟机实例提供块设备，同时备份块设备数据到 Swift 中。
- Sahara 通过 Heat 编排集群配置；在 Swift 中保存数据或二进制文件；将任务分派给虚拟机实例处理；通过 Nova 运行数据处理实例；在 Glance 中注册 Hadoop 镜像。
- Heat 可以编排 Cinder、Neutron、Glance 和 Nova 各种资源。

## 1.4 OpenStack 服务组件

OpenStack 项目通过一系列相互关联的内部服务组件提供了基础设施即服务（Infrastructure as a Service，IaaS）的解决方案，每个内部服务组件都提供了有利于集成的应用程序接口（API）。根据实际需要，可以选择安装某几个或全部服务组件。

下表详细描述了构成 OpenStack 项目的各服务组件功能。

| OpenStack 服务组件 | | |
|---|---|---|
| 服务组件名称 | 代码名称 | 描述 |
| Dashboard（仪表板） | Horizon | 提供基于 Web 与内部各服务组件进行交互的界面。例如，创建虚拟机实例、分配 IP 地址和配置访问控制等 |
| Compute（计算） | Nova | 管理 OpenStack 项目中虚拟机实例计算资源的全生命周期，包括创建、调度和结束 |
| Networking（网络） | Neutron | 为其他服务组件提供网络连接功能，如 Nova 计算服务组件；同时，提供 API 给用户使用；支持多个网络供应商的设备和技术 |
| ObjectStorage（对象存储） | Swift | 负责存储和检索随机非结构化的数据对象；因为多副本和弹性扩展等特点，使其具备高度容错功能；与传统的数据存储目录树形式有区别，对象和文件数据保存在多个驱动器中 |

续表

| OpenStack 服务组件 | | |
|---|---|---|
| 服务组件名称 | 代码名称 | 描述 |
| Block Storage（块存储） | Cinder | 为虚拟机实例提供持久性存储。它的架构简化了块存储设备的创建和管理 |
| Identity（认证） | Keystone | 为 OpenStack 项目中的其他服务组件提供了身份认证和授权功能 |
| Image（镜像） | Glance | 存储和检索虚拟机实例磁盘镜像。在虚拟机实例运行期间，Nova 计算服务组件可以使用此服务 |
| Telemetry（计量） | Ceilometer | 监控和计量 OpenStack 项目中各服务组件的费用、标准、扩展和统计 |
| Orchestration（编排） | Heat | 通过 OpenStack-native REST API 和 CloudFormation-compatible Query API 两个 API，按照 HOT 或 CloudFormation 模板编排各种服务组件使用 |
| Database（数据库） | Trove | 提供高稳定性和可扩展的关系型或非关系型数据库服务 |
| Data Processing Service（数据分析） | Sahara | 通过配置相关参数，如 Hadoop 版本、拓扑和硬件配置，提供和扩展 OpenStack 项目中 Hadoop 集群的能力 |

# 第 2 章 计算（Nova）服务介绍

OpenStack 项目中的 Nova 计算服务组件是非常重要、核心的部分，负责承载和管理云计算系统，与其他组件有密切联系。例如，与 Keystone 身份认证服务组件完成服务认证，与 Glance 镜像服务组件协作生成虚拟机实例，在 Horizon 仪表板服务组件上完成操作等。Nova 计算服务组件支持使用标准的 x86 硬件进行横向扩展。

Nova 计算服务组件包括以下部分。

- nova-api service：负责对终端用户调用 Compute API 的接收和反馈。支持 OpenStack Compute API、the Amazon EC2 API 和其他的 Admin API，运行策略和编排活动。

- nova-api-metadata service：负责接收虚拟机实例对 metadata（元数据）的访问请求。一般部署 nova-network 的多主机模式才会使用 nova-api-metadata service。

- nova-compute service：该服务通过 Hypervisor APIs 创建和终止虚拟机实例。Hypervisor APIs 包括 XenAPI for XenServer/XCP、Libvirt for KVM or QEMU 和 VMwareAPI for VMware。该服务接收消息队列中的信息，执行一系列操作命令，如创建和更新虚拟机实例状态，过程极其复杂。

- nova-scheduler service：该服务接收消息队列中的请求，从而决定虚拟机实例运行在哪个计算节点上。

- nova-conductor module：该服务部署支持横向扩展。出于安全考虑，其不能和 nova-compute service 部署在同一节点上。因为 nova-compute service 不能直接访问数据库，所以需要 nova-conductor module 置于它们中间，与它们交互、通信，传递信息。
- nova-cert module：该服务是一个守护进程，用于 X509 证书的认证服务，通常为 euca-bundle-image（打包好将要上传的镜像）生成证书。仅 EC2 API 需要此 module。
- nova-network worker daemon：类似于 nova-compute service，从消息队列中接收和处理关于网络的请求，如配置桥接网络端口、修改防火墙规则等。
- nova-consoleauth daemon：该进程必须和控制台管理程序（novncproxy、spicehtml5proxy、xvpvncproxy）共同使用，起安全作用。
- nova-novncproxy daemon：通过 VNC 连接 OpenStack 环境中正在运行的虚拟机实例。无须安装 VNC 客户端，可以使用浏览器访问。
- nova-spicehtml5proxy daemon：通过 SPICE 连接 OpenStack 环境中正在运行的虚拟机实例。需要使用 HTML5 访问。
- nova-xvpvncproxy daemon：通过 VNC 连接 OpenStack 环境中正在运行的虚拟机实例。使用 OpenStack 指定的 Java 客户端访问。
- nova-cert daemon：X509 证书。
- euca2ools client：用于管理 Amazon EC2 云资源，不属于 OpenStack 项目。
- nova client：用户通过 nova client 提交请求，包括管理员和非管理员。
- The queue：用于进程和服务间传输信息。支持 RabbitMQ、Apache Qpid 和 ZeroMQ，RabbitMQ 使用频率较高。
- SQL database：用于存储各种信息。只要是 Python 编程语言支持的数据库，Nova 计算服务组件都可以使用。比较常见的数据库包括 SQLite3、MySQL 和 PostgreSQL，MySQL 使用频率较高。

## 2.1 架构设计

Nova 计算服务组件使用基于消息、无共享、松耦合、无状态的架构。OpenStack 项目中的核心服务组件都运行在多台主机节点上,包括 Nova、Cinder、Neutron、Swift 和 Glance 等服务组件,状态信息都存储在数据库中。控制节点服务通过 HTTP 与内部服务进行交互,但与 Scheduler 服务、网络和卷服务的通信依赖高级消息队列协议进行。为避免消息阻塞而造成长时间等待响应,Nova 计算服务组件采用异步调用的机制,当请求被接收后,响应即被触发,发送回执,而不关注该请求是否被完成。

OpenStack 项目中的控制节点服务影响着整个云环境的状态,API 服务器为控制节点服务扮演着 Web 服务的前端,处理各种交互信息。计算节点提供各种计算资源和计算服务。Nova 计算服务组件中的网络服务提供虚拟网络,使实例能够彼此访问和访问公共网络,当然也可以使用 Neutron 网络服务组件代替该功能。目前 Nova 计算服务组件中的网络模块 nova-network 已经过时,代码已经不再更新,强烈建议使用 Neutron 网络服务组件。Scheduler 服务的功能是选择最合适的计算节点运行虚拟机实例。

## 2.2 虚拟化技术介绍

Nova 计算服务组件通过 API 服务器控制虚拟机管理程序。基于预算控制、资源限制、支持特性和所需技术要求等因素,我们需要选择一个最佳的虚拟化技术。根据目前 OpenStack 的发展情况,大部分 Nova 计算服务组件采用的是 KVM 和 Xen 虚拟化技术。

Nova 计算服务组件具备一个抽象层,其允许在部署时选择一种虚拟化技术。现实情况是,每种虚拟化技术的支持和特性是不同的,测试完整度也不相同,并非全部支持相同的特性。以下从测试完整度和功能多样性两方面详细阐述每种虚拟化技术所支持的特性。

### 1. 测试特性

1)虚拟化技术—测试范围

- 测试范围:

➢ 单元模块化测试。

➢ 功能性测试。

- 虚拟化技术一：QEMU/KVM。

2）虚拟化技术二测试范围

- 测试范围：

➢ 单元模块化测试。

➢ 外部组织进行功能性测试。

- 虚拟化技术二：

➢ Hyper-V。

➢ VMware。

➢ XenServer。

➢ 基于 Libvirt 的 Xen。

3）虚拟化技术三测试范围

截至 Icehouse 版本，该组中的虚拟化技术已经过期，不再更新。这些都没有经过充分的测试，使用它们存在一定的风险。

- 测试范围：

➢ 单元模块化不完全测试。

➢ 未公开的功能性测试。

- 虚拟化技术三：

➢ Baremetal（裸金属）。

➢ Docker（容器技术）。

➢ 基于 Libvirt 的 LXC。

### 2. 虚拟化技术支持功能特性

以下基于虚拟化技术支持特性帮助用户选择合适的方案：

- 从 OpenStack 发展轨迹观察，KVM 是最通用的虚拟化技术，在社区中可以找到许多错误的解决方案。KVM 和 QEMU 支持 OpenStack 的所有特性。

- 由于 Microsoft Hyper-V 是免费的，因而得到很多支持。VMware ESXi 由于部分免费，也得到一些支持，但是没有 vCenter 和企业级 License，一些 API 的使用是受限的。

- Xen 技术分类较多，包含以下几种。

    > XenServer：开源，但是有 Citrix 的商业支持。

    > Xen Cloud Platform (XCP)：开源版本，等同于 XenServer。XenServer 支持的特性，XCP 全部支持。

    > XenAPI：管理 XenServer 和 XCP 的 API 程序。

    > XAPI：XenServer 和 XCP 的主守护进程，与 XenAPI 直接通信。

    > 基于 Libvirt 的 Xen：Xen 虚拟化技术使用 Libvirt 驱动。

通过 XenAPI，Nova 计算服务组件支持 XenServer 和 XCP 两种虚拟化技术。但是这并不意味着支持所有基于 Xen 的平台，如运行 SUSE 和 RHEL 的平台，它们都是基于 Libvirt 虚拟化层的 Xen 虚拟化技术。

- 对于 Baremetal 而言，技术已经过期，不再更新，Baremetal 驱动也不再更新。新的特性会被更新到 Ironic 中，由 Ironic 代替 Baremetal。

下面对目前主流的两种虚拟化技术 KVM 和 Xen 进行详细介绍。

## 2.2.1　KVM 虚拟化技术

KVM（Kernel-based Virtual Machine）是基于 x86 硬件虚拟化扩展（Intel VT 或 AMD-V）的全虚拟化解决方案，它包含一个可加载的内核模块 kvm.ko，提供核心的虚拟化基础架构，还有一个处理器特定模块 kvm-intel.ko 或 kvm-amd.ko。

注意:

全虚拟化解决方案:虚拟机与底层硬件之间由一个虚拟化逻辑层 Hypervisor 来完全模拟底层硬件,上层虚拟机完全感知不到运行在虚拟硬件上。优点是虚拟机操作系统内核不需要进行特殊配置,部署便捷、灵活、兼容性好。缺点是虚拟机操作系统的内核不能直接管理底层硬件,内核通过 Hypervisor 管理底层硬件,有转换性能开销。

半虚拟化解决方案:虚拟机操作系统内核需要经过修改,与宿主机操作系统内核共享底层硬件实现。优点是半虚拟化的虚拟机操作系统内核能够直接管理底层硬件,性能表现比全虚拟化技术更好。缺点是虚拟机操作系统内核需要事先进行修改,部署的便捷性、灵活性和兼容性差。

KVM 允许在一台主机节点上运行多个未经改动的虚拟镜像,包括 Windows 和 Linux。每台虚拟机都有独立的虚拟硬件,包括网卡、磁盘等。

KVM 是一个开源项目,其核心组件包含在 Linux 更新主线中(从 Linux 内核 2.6.20 版本开始),用户空间组件包含在 QEMU 更新主线中(从 QEMU 1.3 版本开始)。

KVM 是在 CPU 硬件支持基础之上的虚拟化技术,同 Hyper-V、Xen 一样依赖此项技术。没有 CPU 硬件虚拟化的支持,KVM 是无法工作的。

KVM 是 Linux 的一个模块,可以用 modprobe 加载 KVM 模块。只有在加载模块后,才能进一步通过其他工具创建虚拟机。但仅有 KVM 模块是远远不够的,因为用户无法直接控制内核模块去做事情,必须有一个用户空间的工具才行,这个用户空间的工具就是开源虚拟化软件 QEMU。QEMU 也是一个虚拟化软件,其特点是可虚拟不同的硬件,比如在 x86 的 CPU 上可虚拟一个安腾的 CPU,并可利用它编译出可运行在安腾上的程序。KVM 使用了 QEMU 的一部分,并稍加改造,就变成了可控制 KVM 的用户空间工具。仔细观察,发现官方提供的 KVM 下载有两大部分三个文件,分别是 KVM 模块、QEMU 工具及二者的合集。也就是说,可以只升级 KVM 模块,也可以只升级 QEMU 工具。

一个普通的 Linux 进程有两种运行模式:内核模式和用户模式。但 KVM 拥有 3 种模式,分别是内核模式、用户模式和客户模式。在 KVM 模型中,每个虚拟主机都是由 Linux 调度程序管理的标准进程,该进程调用 KVM 用户模式,执行应用程序。对于应用程序而言,用户模式是默认模式,当需要一些来自内核的服务时便切换到内核模式,如在磁盘上写入数据时。客户模式进程运行在虚拟机内，拥有自己的内核和用户空间变量，在客户模式下可以使用正常的 kill 和 ps 命令，KVM 虚拟机表现为一个正常的进程，可以像其他进程一样被杀掉。KVM 利用硬件虚拟技术模拟处理器的形态，虚拟机的内存管理由内核直接处理。内核模式在需要的时候，向 QEMU 进程发送信号处理大部分的硬件仿真。KVM 管理 CPU 和内存的访问调用，QEMU 仿真硬件资源（如硬盘、网卡、声卡等）。当 QEMU 单独运行时，也可以同时模拟 CPU 和硬件。

KVM 的 API 是通过/dev/kvm 设备访问的，/dev/kvm 是一个标准的字符设备，可以使用常用的 open、close、ioctl 接口操作，但是在 KVM 的实现中，没有提供 read 和 write 接口，所有对 KVM 的操作都是通过 ioctl 接口操作的。KVM 提供给上层的 API 功能可分为 3 种类型。

（1）system 指令：针对虚拟化系统的全局性参数进行设置和控制，包括全局性的参数设置和虚拟机创建等工作，主要指令如下表所示。

| 指　　令 | 说　　明 |
| --- | --- |
| KVM_CREATE_VM | 创建 KVM 虚拟机 |
| KVM_GET_API_VERSION | 查询当前 KVM API 版本 |
| KVM_GET_MSR_INDEX_LIST | 获得 MSR 索引列表 |
| KVM_CHECK_EXTENSION | 检查扩展支持情况 |
| KVM_GET_VCPU_MMAP_SIZE | 运行虚拟主机和用户空间共享内存区域的大小 |

其中，KVM_CREATE_VM 比较重要，用于创建虚拟机，并返回一个代表该虚拟机的文件描述符（fd）。新创建的虚拟机没有虚拟 CPU，也没有内存等资源，需要对新创建虚拟机时返回的文件描述符使用 ioctl 指令进行下一步的配置，生成虚拟 CPU 和内存等资源。

（2）VM 指令：针对虚拟机进行控制，大部分需要针对从 KVM_CREATE_VM 中返回的文件描述符（fd）进行操作，包括配置内存、配置虚拟 CPU、运行虚拟主机等，主要指令如下表所示。

| 指　　令 | 说　　明 |
| --- | --- |
| KVM_CREATE_VCPU | 为虚拟机创建 VCPU |

续表

| 指　令 | 说　明 |
|---|---|
| KVM_RUN | 根据 kvm_run 结构体信息，运行 VM 虚拟机 |
| KVM_CREATE_IRQCHIP | 创建虚拟 APIC，且随后创建的 VCPU 都关联到此 APIC |
| KVM_IRQ_LINE | 对某虚拟 APIC 发出中断信号 |
| KVM_GET_IRQCHIP | 读取 APIC 的中断标志信息 |
| KVM_SET_IRQCHIP | 写入 APIC 的中断标志信息 |
| KVM_GET_DIRTY_LOG | 返回脏内存页的位图 |

　　KVM_CREATE_VCPU 和 KVM_RUN 是 VM ioctl 指令中两种重要的指令，通过 KVM_CREATE_VCPU 为虚拟机创建虚拟 CPU，并获得对应的 fd 文件描述符后，可以对其调用 KVM_RUN，以启动该虚拟机（或称为调度虚拟 CPU）。

　　KVM_RUN 结构体定义在 include /linux/kvm.h 中，可以通过该结构体了解 KVM 的内部运行状态。

　　（3）VCPU 指令：针对具体的虚拟 CPU 进行参数设置，包括寄存器读/写、中断设置、内存设置、时钟管理、调试开关等。KVM 虚拟机运行时也可以进行相关设置。主要指令如下。

- 寄存器控制方面的主要指令如下表所示。

| 指　令 | 说　明 |
|---|---|
| KVM_GET_REGS | 获取通用寄存器信息 |
| KVM_SET_REGS | 设置通用寄存器信息 |
| KVM_GET_SREGS | 获取特殊寄存器信息 |
| KVM_SET_SREG | 设置特殊寄存器信息 |
| KVM_GET_MSRS | 获取 MSR 寄存器信息 |
| KVM_SET_MSRS | 设置 MSR 寄存器信息 |
| KVM_GET_FPU | 获取浮点寄存器信息 |
| KVM_SET_FPU | 设置浮点寄存器信息 |

| 指　　令 | 说　　明 |
| --- | --- |
| KVM_GET_XSAVE | 获取虚拟 CPU 的 XSAVE 寄存器信息 |
| KVM_SET_XSAVE | 设置虚拟 CPU 的 XSAVE 寄存器信息 |
| KVM_GET_XCRS | 获取虚拟 CPU 的 XCR 寄存器信息 |
| KVM_SET_XCRS | 设置虚拟 CPU 的 XCR 寄存器信息 |

- 中断和事件管理方面的主要指令如下表所示。

| 指　　令 | 说　　明 |
| --- | --- |
| KVM_INTERRUPT | 在虚拟 CPU 上产生中断（当 APIC 无效时） |
| KVM_SET_SIGNAL_MASK | 设置某个虚拟 CPU 的中断信号屏蔽掩码 |
| KVM_GET_CPU_EVENTS | 获取虚拟 CPU 中断被挂起待延时处理的事件，如中断、NMI 或异常等 |
| KVM_SET_CPU_EVENTS | 设置虚拟 CPU 的事件，如中断、NMI 或异常等 |

- 内存管理方面的主要指令如下表所示。

| 指　　令 | 说　　明 |
| --- | --- |
| KVM_TRANSLATE | 将虚拟 CPU 的物理地址翻译成 HPA |
| KVM_SET_USER_MEMORY_REGION | 修改虚拟 CPU 的内存区域 |
| KVM_SET_TSS_ADDR | 在 Intel 架构下初始化 TSS 内存区域 |
| KVM_SET_IDENTITY_MAP_ADDR | 在 Intel 架构下创建 EPT 页表 |

- 其他方面的主要指令包括 CPUID 设置、调试接口等。

对于 KVM 的操作都是从打开 /dev/kvm 设备文件开始的，打开后，会获得相应的文件描述符（fd），然后通过 ioctl 系统指令对该 fd 进行进一步的操作，如通过 KVM_CREATE_VM 指令可以创建一个虚拟机并返回虚拟机对应的文件描述符，接着根据该描述符来进一步控制虚拟机的行为，如通过 KVM_CREATE_VCPU 指令来为该虚拟机创建 VCPU。下面通过图示介绍 KVM 的初始化过程。

1．加载模块

加载 KVM 和 SVM 或 VMX。成功加载后，生成/dev/kvm 文件供用户空间访问，如下图所示。

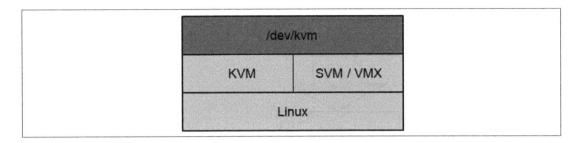

2．调用/dev/kvm

/dev/kvm 既不可读也不可写。

/dev/kvm 拥有 ioctl 接口，如下图所示。

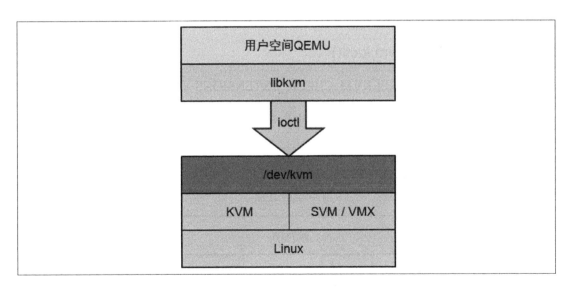

3．调用kvm ioctl()

调用 KVM_GET_API_VERSION 进行版本检查。

调用 KVM_CREATE_VM 创建一个虚拟主机，并返回一个 kvm-vm 文件描述符，如下图所示。

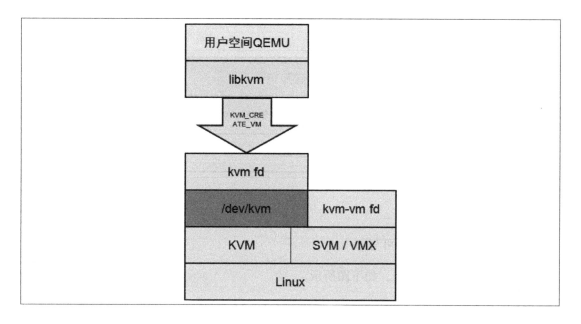

### 4．调用kvm和kvm-vm ioctl()

在 kvm 文件描述符上调用 KVM_CHECK_EXTENSIONS（检查扩展支持情况）。

在 kvm-vm 文件描述符上调用 KVM_SET_TSS_ADDR（在 Intel 架构中初始化 TSS 内存区域）设置 TSS 地址，如下图所示。

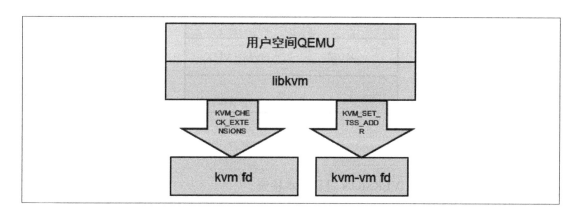

## 5．调用kvm和kvm-vm ioctl()

在 kvm 文件描述符上调用 KVM_CHECK_EXTENSIONS（检查扩展支持情况）。

在 kvm-vm 文件描述符上设置 KVM_SET_MEMORY_REGION（修改虚拟 CPU 的内存区域），如下图所示。

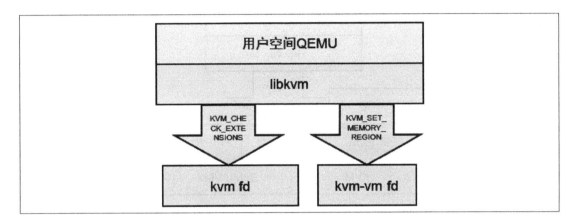

## 6．调用kvm和kvm-vm ioctl()

在 kvm 文件描述符上调用 KVM_CHECK_EXTENSIONS（检查扩展支持情况）。

在 kvm-vm 文件描述符上调用 KVM_CREATE_IRQCHIP（创建虚拟 APIC 并将虚拟 CPU 关联到 APIC）设置 irqchip，如下图所示。

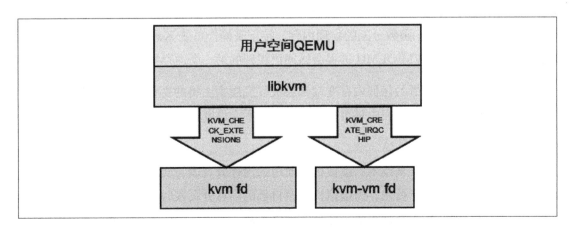

## 7. 创建虚拟CPU

在 kvm 文件描述符上调用 KVM_CREATE_VCPU。

执行后，返回 kvm-vcpu 文件描述符，如下图所示。

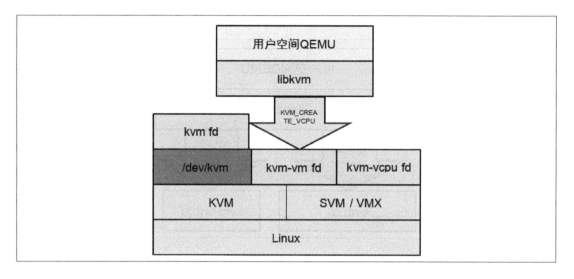

KVM/QEMU 的内存管理非常有趣，因为 KVM/QEMU 运行就像 Linux 系统中的一个程序运行，所以它分配内存是调用 malloc() 和 mmap()函数进行的。当一个虚拟机申请 1GB 的物理内存时，KVM/QEMU 会执行 malloc(1<<30) 操作，从宿主机上分配 1GB 的虚拟地址空间。所以它只调用 malloc()函数，并没有进行实际的物理内存分配；而是当虚拟机第一次启动需要访问内存时，才会给虚拟机分配真正的物理内存。虚拟机操作系统启动运行，它可以识别通过 malloc()函数分配的物理内存，接下来操作系统 Kernel 开始访问已识别的物理内存地址，这时 KVM/QEMU 进程会访问已识别的第一个内存页。

KVM/QEMU 虚拟机的任何内存变动都会关联底层宿主机的变化，宿主机会确认该虚拟机变化在其整个内存分页表中是否有效、可用，不允许其访问任何不属于它的内存页。此内存运行访问机制有两种。

- 第一种是影子分页表技术。虚拟机所使用的内存分页表与实际的内存分页表是独立的，不是同一张分页表。当虚拟机修改自己的内存分页表时，宿主机会检测到有修改动作发生，然后进行确认，之后才会修改真正的分区表，使由虚拟机发起的修改操作生效。虚拟机不能直接访问真正的内存分页表，而是访问影子分页表。这是一种非常普通的虚拟化内存技术。

- 第二种是 VMX/AMD-V 扩展技术。VMX/AMD-V 扩展技术允许底层宿主机始终监控，以此获得虚拟机修改真正内存分页表的信息。这种内存运行访问机制实际且有效，但是它对性能有一些影响，完成一次访问最高可能消耗 25 页内存，代价非常大。造成这种问题的原因是每次内存访问需要两次操作才能完成，包括虚拟机内存分页表访问和宿主机内存分页表访问。当然，宿主机运行和维护影子分页表也需要消耗大量资源。

AMD 和 Intel 厂商为了解决这些性能损耗问题，开发了两种全新的技术 EPT 和 NPT。这两种技术很相似，都使硬件重新识别架构，快速地将虚拟机内存变化直接传递给宿主机的物理内存，而不用再去访问宿主机内存分页表，减少一次操作，效率更高。

但由此带来的问题就是宿主机的内存分页表像进程隔离一样会被强制执行，当一个内存页在宿主机内标记为没有被分配时，实际上该内存页可能已经被虚拟机占有了，所以必须与 EPT/NPT 共同协商处理这个变化。

为了解决该问题，可以在软件层面通过调用 mmu_notifiers() 函数解决此问题，因为 KVM/QEMU 的内存本来就是正常的物理内存，Kernel 可以像交换、替换和释放正常的物理内存一样处理这些内存页。

在虚拟机释放内存页给宿主机前，需要等待宿主机的通知。当 KVM/QEMU 虚拟主机在影子分页表或 EPT/NPT 中删除该内存页后，宿主机便可以自由地使用该内存页。

关于内存的申请与回收，可以总结为以下几个步骤。

（1）内存申请。

① QEMU 调用 malloc() 函数为虚拟机分配虚拟内存页，但是此时并没有申请真正的物理内存。

② 虚拟机开始访问该虚拟内存页，并且认为该虚拟内存页是真正的物理内存页，但是由于该内存页没有被真正分配，所以开始向宿主机申请。

③ 宿主机内核发现有一个内存页错误，便会在已经被分配的 malloc()'d 区域调用

do_page_fault()函数。如果一切顺利，没有打断，则宿主机开始响应虚拟机的操作。

④ 宿主机内核创建 pte_t，使 malloc()'d 虚拟地址连接到真正的物理内存地址，生成 rmap，并把它们放到 LRU 中。

⑤ 此时，mmu_notifier_change_pte()被调用，其允许 KVM 为该内存页创建 NPT/EPT。

⑥ 宿主机从该错误的内存页中返回标识，虚拟机得到内存后执行操作恢复。

（2）内存回收。

① 宿主机内核利用 rmap 结构寻找到需回收的内存页被映射到哪个 VMA（vm_area_struct）。

② 宿主机内核查找该 VMA 所关联的 mm_struct，并遍历宿主机的内存分页表，查找到该内存页在物理硬件上的位置。

③ 宿主机内核替换出该内存页并清空 pte_t。

④ 宿主机内核接着调用 mmu_notifier invalidate_page()函数，在 NPT/EPT 中查找到该页并删除。

⑤ 现在，该页已经被释放，任何需要该页的访问都可以向宿主机申请（此时，可以转换到内存申请的第二步）。

## 2.2.2 Xen 虚拟化技术

Xen 是一种开源的、属于类型 1（裸金属虚拟化，Baremetal Hypervisor）的虚拟化技术，它使多个同样操作系统或不同操作系统的虚拟机运行在同一个物理主机节点上成为可能并实现。Xen 是唯一的属于类型 1（裸金属虚拟化，Baremetal Hypervisor）并且开源的虚拟化技术，它被作为商业应用或开源应用的基础而加以使用，如服务器虚拟化、Infrastructure as a Service（IaaS）、桌面虚拟化、安全应用、嵌入式和硬件设备等。由于它性能稳定，因而被广泛用于云计算生产环境。

以下是 Xen 虚拟化技术的一些关键特性。

- 轻便小型的设计（核心代码有 1MB 左右）。它使用了微小内核设计，占用极少内存，加上有限的接口设计，使得它比其他虚拟化技术更健壮、更安全。
- 操作系统无关性。Domain0 一般安装在 Linux 操作系统中，也可以使用其他操作系统代替，如 NetBSD 和 OpenSolaris 等。
- 驱动分离。Xen 虚拟化技术允许主要的硬件设备驱动运行于虚拟机内部，当驱动出现 crash（宕机）或者报错时，包含该驱动的虚拟机可以重启，该驱动也可以重启，这样不会影响其他的虚拟机。
- 半虚拟化技术。运行在半虚拟化技术上的虚拟机已经经过优化，它们可以运行得更加流畅，比运行在需要硬件扩展支持的全虚拟化管理程序（HVM）上的虚拟机更快。当然，Xen 可以运行在不支持硬件扩展的硬件平台上。

Xen 虚拟化技术架构包含 3 个关键点，掌握这 3 个关键点对于用户理解和做出正确的选择至关重要。

- 类型：Xen 虚拟机技术支持半虚拟化（Para-Virtualization，PV）和全虚拟化（Hardware-assisted Virtualization，HVM）两种类型。
- Domain0：Xen 虚拟化架构中包含一个特殊的域 Domain0，其包括硬件设备驱动和控制虚拟机的 Toolstack。
- Toolstacks：其涵盖各种不同的 Toolstack。

**注意：**

类型 1（裸金属虚拟化，Baremetal Hypervisor）：该类型的虚拟化技术直接运行在物理主机节点硬件设备上，并且管理虚拟机操作系统，如下图所示。第一个此类型的虚拟化技术是在 1960 年由 IBM 发布的，它包括一个测试软件 SIMMON 和 CP/CMS 操作系统（IBM 的 z/VM 操作系统的前身）。目前比较流行的产品或技术包括 Oracle VM Server for SPARC、Oracle VM Server for x86、Citrix XenServer、Microsoft Hyper-V 和 VMware ESX/ESXi。

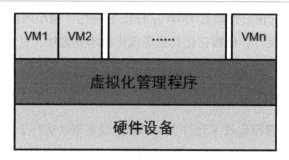

类型1

类型 2（可托管的虚拟化，Hosted Hypervisor）：该类型的虚拟化技术允许运行在普通的操作系统上，和普通的计算机程序类似，如下图所示。一个运行的虚拟机作为一个进程存在于物理主机节点的操作系统中。目前比较流行的产品或技术包括 VMware Workstation、VMware Player、VirtualBox 和 QEMU。

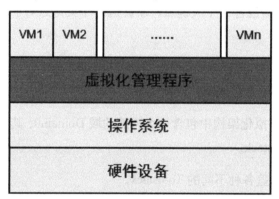

类型2

尽管对虚拟化技术的架构进行了分类，并且划分了类型 1 和类型 2，但是在实际运用和当今 IT 技术发展中，它们彼此之间并没有如此严格的分类。Linux 的 Kernel-based Virtual Machine（KVM）和 FreeBSD 的 BHyVe 是基于内核的虚拟化技术，利用该基于内核的虚拟化技术将传统的操作系统转化成类型 1 的虚拟化管理程序，与此同时，Linux 发行版和 FreeBSD 仍然是通用的操作系统，与其他应用一起竞相争夺虚拟机操作系统可以使用的资源。所以据此分析，KVM 和 BHyVe 虚拟化技术属于类型 2 的虚拟化技术架构。

## 1．Xen 架构

Xen 虚拟化管理程序直接运行在硬件之上，处理各种 CPU、内存和中断请求。在包含虚拟化管理程序的操作系统启动过程中，BootLoader 加载完成并退出后，加载的第一个程序就是 Xen，在其上运行着虚拟机。运行的虚拟机叫作 Domain（域）或客户机（Guest），其中有一个特殊的 Domain 叫作 Domain0，其包含了所有硬件设备的驱动。Domain0 还包含了控制栈（Toolstack），用于虚拟机创建、删除和配置等。下图是 Xen 虚拟化管理程序的架构。

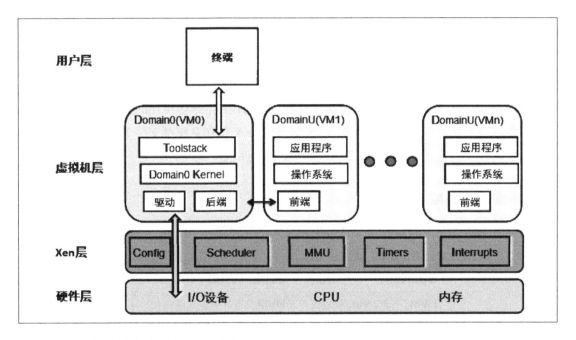

在 Xen 架构中包含以下几个重点。

- Xen 虚拟化管理程序是一个极小的软件程序，包含大概 15 万行代码。Xen 虚拟化管理程序本身没有 I/O 功能虚拟化，如网络和存储等。
- 虚拟机是一个虚拟化的环境，每个虚拟机都运行着自己的操作系统和应用程序。Xen 支持两种虚拟化模式：半虚拟化和全虚拟化。在同一个虚拟化管理程序上可以同时并行使用这两种虚拟化模式，也可以串行在全虚拟化模式上使用半虚拟化模式，以此保证半虚拟化和全虚拟化的连续性。虚拟机与硬件之间是完全隔离的，它们没有任何权限可以访问底层的硬件和 I/O 设备等，因此它们也被叫作 DomainU（Unprivileged Domain）。

- Domain0 是一个特殊的虚拟机，其具备特殊的、足够的权限直接访问底层的硬件设备，处理所有底层的 I/O 设备请求，并与其他虚拟机（DomainU）进行交互通信。Domain0 对外部开放一个接口，使用户可以控制整个系统。没有 Domain0，Xen 虚拟化管理程序是无法使用的，它是整个系统启动后加载的第一个虚拟机。
- Toolstack 包含在 Domain0 中，也叫作控制栈，其允许用户管理虚拟机，包括虚拟机创建、删除和配置等。
- 终端是 Toolstack 对外部开放的一个接口，用户可以通过命令行或/和图形化界面控制整个系统，OpenStack 和 CloudStack 中的编排服务也被支持。
- Domain0 要求一个支持 Xen 虚拟化管理程序的内核，半虚拟化的虚拟机（DomainU）要求一个支持半虚拟化的内核。比较新的 Linux 操作系统都支持 Xen 虚拟化管理程序，并且也包含支持虚拟化和 Toolstack 的软件包。

### 2．虚拟机类型

下图是 Xen 虚拟化管理程序支持虚拟化模式的变化。

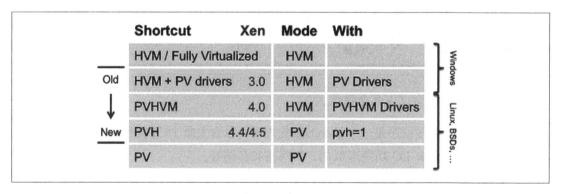

1）半虚拟化（PV）

半虚拟化是由 Xen 虚拟化管理程序引入的一个轻量级、高效的虚拟化模式，之后被其他虚拟化平台所采用。半虚拟化不要求物理主机节点 CPU 具备扩展特性，但是其需要支持半虚拟化的内核和驱动，因此，虚拟机能够感知到虚拟化管理程序；同时，因为没有硬件仿真，所以运行非常高效。支持半虚拟化的内核包括 Linux、FreeBSD、NetBSD 和 OpenSolaris。Linux 内核从 2.6.24 版本开始，使用 Linux pvops framework 的内核都支持半虚拟化，这也就意味着除了比较老的版本外，几乎所有的 Linux 内核都支持半虚拟化。下图是半虚拟化模式在 Xen 虚拟化管理程序中的性能表现。

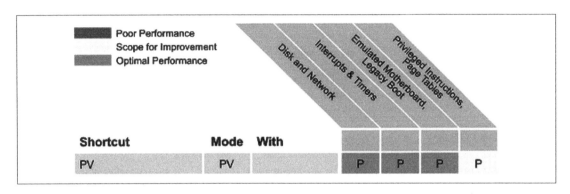

2）全虚拟化（HVM）

全虚拟化需要物理主机节点 CPU 扩展特性的支持，为此，Intel 和 AMD 厂商提供了 Intel VT 和 AMD-V 技术。Xen 虚拟化管理程序使用 QEMU 仿真硬件设备，包括 BIOS、IDE 磁盘控制器、VGA 图形适配器、USB 控制器和网络适配器等。硬件的扩展特性提高了仿真性能。同时，全虚拟化模式下的虚拟机不再需要特殊内核的支持，这也就意味着 Windows 操作系统在基于 Xen 全虚拟化的平台上也是被支持的。一般情况下，半虚拟化的虚拟机比全虚拟化的虚拟机性能表现更好，因为全虚拟化的虚拟机需要硬件仿真，会消耗一部分性能。

在某些情况下，可以使用半虚拟化驱动加速全虚拟化虚拟机的 I/O 性能。在 Windows 虚拟机中，需要安装合适的半虚拟化驱动。具体信息可以参照以下链接。

- Xen 半虚拟化驱动列表：http://xenproject.org/downloads/windows-pv-drivers.html。
- 第三方 GPL 半虚拟化驱动列表：http://wiki.xensource.com/wiki/Xen_Windows_GplPv。
- Windows 半虚拟化驱动列表：http://wiki.xensource.com/wiki/Category:Windows_PV_Drivers。

在 Xen 虚拟化支持的操作系统中，在选择全虚拟化模式运行操作系统时，其已安装的半虚拟化或全虚拟化驱动可以自动被使用。下图展示了全虚拟化模式和含有半虚拟化驱动的全虚拟化模式之间的区别。

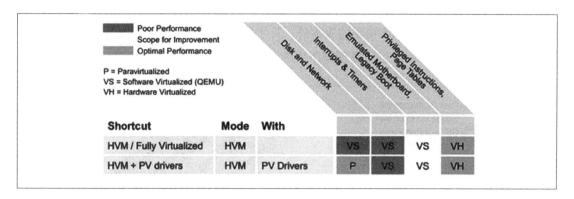

3）全虚拟化模式+半虚拟化驱动（PVHVM）

全虚拟化模式下的虚拟机可以使用指定的半虚拟化驱动，以此达到增强系统性能的目的。这些驱动是为全虚拟化环境而优化的半虚拟化驱动，绕过磁盘和网络的模拟仿真，从而在全虚拟化模式下获得更好的性能。这也就意味着在使用一些虚拟机操作系统时会获得更好的性能，如 Windows 等。

基于 Xen 虚拟化管理程序的半虚拟化虚拟机可以自动使用半虚拟化驱动，全虚拟化模式下使用的半虚拟化驱动仅适用于全虚拟化模式下的虚拟机。下图展示了 3 种类型的全虚拟化模式之间的区别。

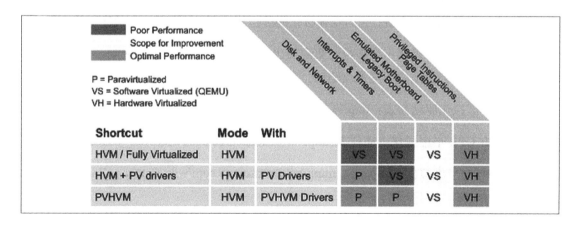

4）半虚拟化+硬件扩展特性（PVH）

Xen 虚拟化管理程序 4.4 版本中包含一种虚拟化模式，叫作基于 DomainU 的 PVH；4.5 版本又开发了一种基于 Domain0（Linux 和 BSD）的 PVH 虚拟化模式，其实质是半虚拟化的虚拟机可以使用半虚拟化驱动以提高 I/O 性能，也可以使用硬件扩展特性提高系统性能，不需要硬件仿真。PVH 在 4.4 和 4.5 版本中作为试验进行发布和测试使用，性能表现非常好，并且在 4.6 版本中进行了优化。从本质上讲，PVH 对两种虚拟化模式进行了合并，简化了 Xen 虚拟化管理程序的架构。

简而言之，PVH 在 Linux 和 BSD 中使用了极少的代码和接口，从而减少了 TCB 和攻击的可能性，降低了风险。一旦对其进行相应的优化，它将具备更好的性能和更低的延迟，特别是在 64 位的操作系统上表现更优。

PVH 要求虚拟机操作系统对其提供支持，在配置文件中设置 pvh=1 即可启用 PVH 支持。下图展示了全虚拟化、半虚拟化和 PVH 之间的区别。

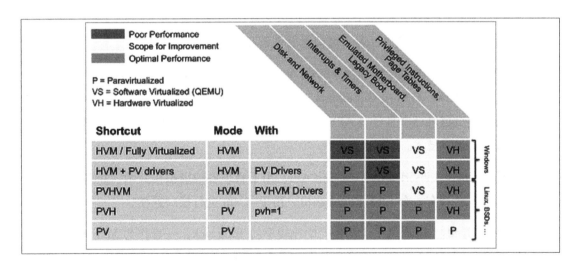

### 3．Toolstacks、APIs和终端

Xen 虚拟化管理程序包含许多不同的 Toolstacks，每个 Toolstack 对外开放接口，利用该开放接口可以运行各种不同的工具，管理整个系统。下面介绍一些商业化产品所使用的 Toolstack，以及托管服务商使用 API 的案例。

Xen 虚拟化管理程序与默认的 Toolstacks、Libvirt 和 XAPI 相互作用、协同运行。Xen 虚拟化管理程序与 XAPI 配合使用的虚拟化产品是 XCP，但是其已经逐渐被开源产品 XenServer 所取代。基于 Xen 虚拟化管理程序的各种虚拟化模式都具备各自的优势，并且针对不同的案例进行了优化，因此对虚拟化方案的选择也是见仁见智。下图中的 Default:XL 和 XAPI/XE 由 Xen 虚拟化管理程序提供。

下面针对图形中各种不同的 Toolstacks 进行解释。

- Default:XL：XL 是一个使用 Libxenlight 建立的轻量级 CLI Toolstack，其随着 Xen 的更新而发展，在 Xen 4.1 版本中，XL 是默认使用的 Toolstack。由于 Xend 已经过时，即将从 Xen 中删除，所以 XL 被设计为与 XM CLI 向后兼容，它提供了一个简单的、针对 Xen 的命令行接口，用于虚拟机的创建和管理等。

- XAPI/XE：Xen 虚拟化管理程序管理 API（XAPI）是 XenServer 虚拟化产品默认使用的 Toolstack，有时在 CloudStack 中也会使用。目前，逐渐被废弃的 XCP 虚拟化平台正尝试提供一个基于 XAPI 的社区开发平台，以开源的形式进行，并在正式发行版中提供该技术。

- Libvirt/virsh：Libvirt 是一个虚拟化 API，用于管理各种虚拟化技术，如 Xen、KVM 等。Libvirt 拥有一个 Libxenlight 端口，用于与 xm 进行接口通信。

- Default/Xend：Xend 是一个诞生时间比较早的 Toolstack，一直作为 Xen 虚拟化管理程序的一部分进行更新、发布和使用。但是自 Xen 4.1 版本发布后，Xend 开始过时，不再推荐使用；在 Xen 4.5 版本以后，其已经被删除。现在推荐使用上述三种 Toolstacks。XL 被设计成一个与 XM CLI 兼容的命令行 Toolstack，可以将请求发送给 Xend，所以使用 XL 不失为一种很好的升级方法。

下表是以上介绍的 Toolstack 与其 CLI 命令行的对应关系。

| Toolstack | XL | XAPI | Libvirt | Xend |
|---|---|---|---|---|
| CLI tool | xl | xe | virsh | xm |

## 2.3　Libvirt 技术介绍

Libvirt 是一个函数库，包含实现 Linux 虚拟化功能的 Linux API，提供了管理虚拟机的通用的、稳定的、统一的接口。其主要包括 Libvirt API、Libvirtd 进程和 virsh 工具集三部分。

为了便于理解，我们将 Libvirt 分为三层，从三个层级进行介绍，具体如下图所示。

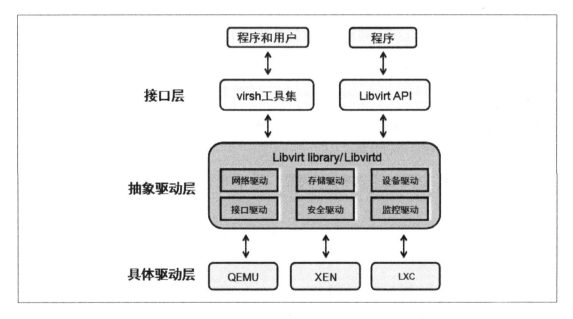

其中，接口层的 virsh 工具集和 Libvirt API 负责接收程序和用户的指令；在抽象驱动层，接收上层发出的指令调用 Libvirt 库或控制 Libvirtd 进程提供统一的接口；在具体驱动层，调用底层相应的虚拟化技术接口执行操作。

Libvirt 可以提供关于虚拟机管理的所有 API，包括创建、修改、监控、控制、迁移和停止虚拟机。当然，并不是所有的虚拟化技术都支持 Libvirt 提供这些关于虚拟机的操作，但是 Libvirt 可以提供关于这些操作的 API。任何一个对虚拟机有价值的操作，Libvirt 都会提供。利用 Libvirt 可以同时访问多个物理主机，但是对单主机节点的操作是有限制的。Libvirt API 提供了管理虚拟机所需的物理主机节点层面的所有操作，如防火墙规则、存储管理和一般配置 API 等。同时，它也提供了实施管理策略所需的状态监控 API，对虚拟机的状态监控也从某种角度反映出物理主机的资源消耗情况，这对掌握生产环境中硬件资源的使用率极其重要。

综上所述，Libvirt 实现了以下功能：

- 所有 API 通过安全协议都可以远端执行操作。
- 大部分 API 对虚拟化和物理主机操作系统的管理都是通用的，但是一些 API 只能应用于虚拟化管理。
- Libvirt API 可以完成虚拟机需要的所有操作。
- Libvirt API 不提供高级别虚拟化策略或多节点管理功能，如负载均衡功能。但是 Libvirt 不拒绝在其外部通过技术实现这些高级功能。
- 目前 Libvirt 的稳定性存在一些问题，建议将经常需要变化的、稳定性低的应用与稳定性要求高的应用进行分离。
- 被管理的物理主机节点可以来自不同的厂商，但是 Libvirt 对其进行远程管理必须使用安全协议。
- Libvirt API 可以监控和使用被管理物理主机节点的资源，包括 CPU、内存、存储和网络等。

由此可以总结出：Libvirt 的目标是建立一套集中在虚拟化层面、面向应用的高级管理工具。

## 2.3.1 Libvirt API 介绍

如前所述，Libvirt API 的功能是使用有效的硬件资源服务当前的虚拟机操作系统。下面从 Libvirt 驱动和 Libvirtd 进程两方面展开介绍。

## 1．Libvirt驱动

Libvirt 驱动是实现 Libvirt 功能的基本模块，保证了 Libvirt 程序可以处理和调用指定的虚拟化程序。Libvirt 驱动在连接处理过程中被发现和注册，每个驱动都有一个注册 API，其负责加载指定的驱动。下图是关于虚拟化程序驱动的简单视图。

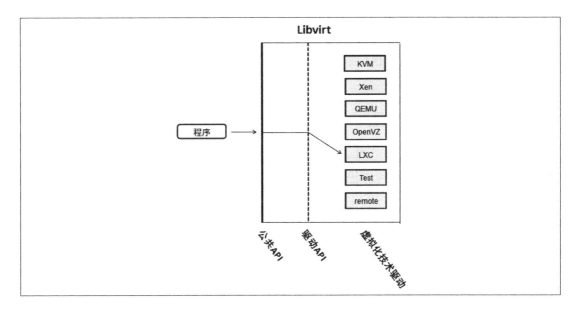

Libvirt 目前支持的虚拟化技术如下表所示。

| 虚拟化技术名称 | 描述 |
| --- | --- |
| Xen | 面向 IA-32、IA-64 和 PowerPC 970 架构的虚拟机监控程序 |
| QEMU | 面向各种架构的平台仿真器 |
| KVM | Linux 平台仿真器 |
| LXC | 用于操作系统虚拟化的 Linux（轻量级）容器 |
| OpenVZ | 基于 Linux 内核的操作系统级虚拟化 |
| VirtualBox | x86 虚拟化虚拟机监控程序 |
| User Mode Linux | 面向各种架构的 Linux 平台仿真器 |
| Test | 面向伪虚拟机监控程序的测试驱动器 |
| Storage | 存储池驱动器（本地磁盘、网络磁盘、iSCSI 卷） |

### 2．Libvirtd进程

Libvirtd进程通过remote驱动可以远程处理对Libvirt驱动的访问请求。一些虚拟化技术支持来自客户端的连接和反馈，如Test、OpenVZ、VMware、PowerVM、VirtualBox、ESX、Hyper-V和Xen等。Libvirtd进程随操作系统启动而启动运行，与普通进程一样，具有足够权限的用户可以重启和停止它。确定需要使用的驱动后，Libvirtd进程可以将请求路由到正确的驱动上，建立连接并检索所请求的信息，返回状态信息或数据给应用。应用可以根据需求确定利用这些数据执行哪些操作，如展示、打印日志等。下图是Libvirtd进程的简单视图。

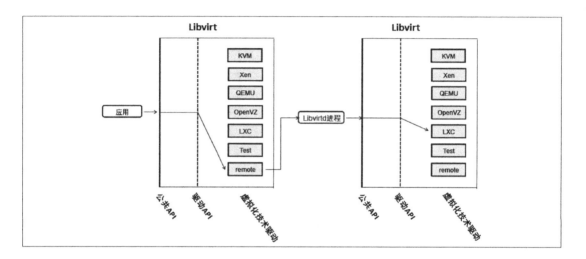

## 2.3.2 Libvirt 网络架构

下面从网络逻辑架构和物理架构两方面来阐述Libvirt的网络设计。

- vlan1：该虚拟网络的流量连接到物理网络2上。
- vlan2：该虚拟网络与物理网络完全隔离。
- 虚拟机A：网络端口eth0桥接到物理网络1上，网络端口eth1连接到vlan1上。
- 虚拟机B：网络端口eth0连接到vlan1上，网络端口eth1连接到vlan2上。该虚拟主机扮演路由的角色，在两个vlan之间可以转发网络包，使虚拟机C连接到物理网络2上。

- 虚拟机 C：仅有的网络端口 eth0 连接到 vlan2 上，没有直接连接物理网络，依靠虚拟机 B 的路由转发流量实现网络互通。

其逻辑架构如下图所示。

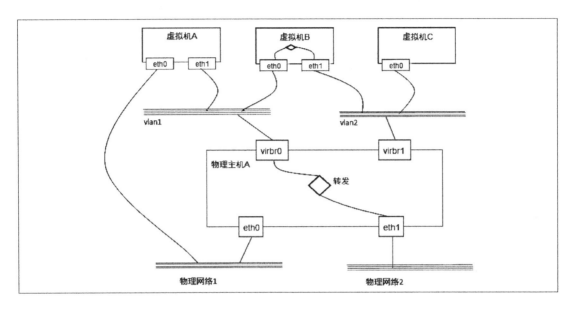

其物理架构如下图所示。

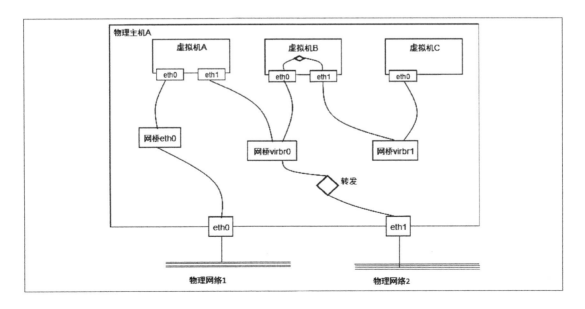

## 2.3.3　Libvirt 存储架构

Libvirt 的存储设计包括两个核心部分。

- 存储卷：一个存储卷可以分配给虚拟机使用，或创建成存储池以供使用。一个存储卷可以是一个块设备、一个.raw 文件或其他格式的文件。
- 存储池：存储池可以理解成将存储资源池化，分割成存储卷分配给虚拟机使用。存储池用来管理物理磁盘、NFS 服务器或一个 LVM 组。

# 第 3 章
# 网络（Neutron）服务介绍

OpenStack 项目中的 Neutron 网络服务组件提供虚拟机实例对网络的连接，其中 plug-ins 能够提供对多种网络设备和软件的支持，使 OpenStack 环境的构建和部署具备更多的灵活性，其最主要的功能是为虚拟机实例提供网络连接。

Neutron 网络服务组件包括以下几部分。

- neutron-server：接收和路由 API 请求到 OpenStack 中的网络 plug-in。

- OpenStack Networking plug-ins and agents：创建端口（Ports）、网络（Networks）和子网（Subnets），提供 IP 地址。plug-ins 和 agents 根据不同的厂商和技术而应用于不同的云环境中，OpenStack 中的 plug-ins 一般支持 Cisco Virtual and Physical Switches、NEC OpenFlow Products、Open vSwitch、Linux Bridging 和 VMware NSX Product。常见的 agents 包括 L3（Layer 3）、DHCP（Dynamic Host Configuration Protocol）和 plug-in agent。

- Messaging queue：在 neutron-server 和 agents 之间路由信息，同时也会作为一个数据库存储 plug-ins 的网络连接状态。

## 3.1 网络 OSI 7 层模型

Open Systems Interconnection（OSI）模型的全称是开放系统互联参考模型，是一个逻辑上的定义，是一个规范，使通信和计算系统自由互联，而不依赖其他内在的架构或技术。它诞生的目标就是使用标准协议使不同平台的系统互联互通。该模型为了更好地被理解和应用，抽象成层的概念，共分为 7 层，每一层既服务于上层，又被下层所服务。7 层由低到高分别是物理层、数据链路层、网络层、传输层、会话层、表示层和应用层。处于同一层的两个实例是互相可见的，并且是水平连接的。

OSI 模型属于 International Organization Standardization（ISO）项目，由 ISO/IEC 7498-1 进行维护。

OSI 模型最大的优点是将服务、接口和协议明确分开，其中，服务说明某层为上一层提供某些功能，接口说明上层如何调用下层所提供的服务，协议则负责如何实现本层服务。因此，OSI 模型中的每一层都具备很强的独立性，互相连接的网络中各层所采用的协议是没有限制的，只要向上层提供服务并且不改变接口即可。

下表是对 OSI 模型 7 个层级的描述。

| OSI 模型 | | | | |
|---|---|---|---|---|
| 层 | 级 | 协议数据单元（PDU） | 功　　能 | 举　　例 |
| 主机层 | 7. 应用层 | 数据 | 高级 APIs 调用，包括资源共享、远程文件访问、目录服务和虚拟终端 | NFS,SMB,AFP,FTAM,NCP |
| | 6. 表示层 | | 网络服务与应用间的数据传输，包括字节编码、数据压缩和加密/解密 | CSS,GIF,HTML,XML,JSON,S/MIME |
| | 5. 会话层 | | 管理通信会话，如在两个节点间多次传输的持续信息交换 | RPC,SCP,PAP,TLS,FTP,HTTP,HTTPS,SMTP,SSH,Telnet |
| | 4. 传输层 | TCP/UDP | 同一网络中点对点的可靠数据段传输，包括切割、确认和复用 | NBF,TCP,UDP |
| 介质层 | 3. 网络层 | 包（Package） | 构建和管理多节点间网络，包括地址、路由和流量的控制 | AppleTalk,ICMP,IPSec,IPv4,IPv6 |

续表

| OSI 模型 |||||
| --- | --- | --- | --- | --- |
| 层 级 || 协议数据单元（PDU） | 功 能 | 举 例 |
| 介质层 | 2. 数据链路层 | 帧（Frame） | 在同一层互相连接的两个节点间进行数据帧的可靠传输 | IEEE802.2,L2TP,LLDP,MAC,PPP,ATP,MPLS |
| | 1. 物理层 | 比特（Bit） | 通过物理介质传输和接收原始比特流 | DOCSIS,DSL,Ethernet Physical Layer,ISDN,RS-232 |

OSI 模型中每个层级相互通信设备上的两个实体，在相同层级上使用协议交换 Protocol Data Unit（PDU）。每个 PDU 中包含的有效数据叫作 Service Data Unit（SDU），其包含每层协议相关的数据头和数据尾信息。在 OSI 模型中，两个相互通信的设备之间的数据处理过程如下：

（1）一个拥有 $N$ 层的传输设备在其最高层级将要传输的数据解析成 Protocol Data Unit（PDU）。

（2）PDU 传输到 $N$-1 层，在该层级，此 PDU 作为 Service Data Unit（SDU）被识别。

（3）在 $N$-1 层，该 SDU 使用数据头、数据尾或二者共同使用进行连接、组合，生成全新的 $N$-1 层的 PDU，然后继续将其传递给 $N$-2 层。

（4）拆分和组合的流程和动作会一直重复，直至数据到达底层，也就是数据从发送设备传递到目标设备。

（5）在目标设备上，数据以 SDU 格式从底层传递到最高层。在数据传递过程中，不断地剥离每层的数据头、数据尾，直至数据到达最高层，也就是数据最终被应用、消耗的地方。

下面对 OSI 模型进行详细介绍。

### 1．物理层

物理层并不是物理媒介本身，它只是开放系统中利用物理媒介实现物理连接的功能描述和执行连接的规程。物理层提供用于建立、保持和断开物理连接的机械的、电气的、功能的和过程的条件。简而言之，物理层提供有关同步和全双工比特流在物理媒介上的传输手段。物理层有以下主要功能：

- 定义数据连接电气和物理规格，为设备之间的数据通信提供传输媒介及互连设备，为数据传输提供可靠的环境。物理层的媒介包括架空明线、平衡电缆、光纤、无线设备（5 GHz 或 2.4 GHz）等。
- 负责传输和接收非结构化原始数据。
- 定义传输模式，如单工模式、半双工模式和全双工模式。
- 定义网络拓扑类型，如总线、网格和环等。
- 主要处理原始数据。

### 2. 数据链路层

数据链路层为互相连接的节点提供数据传输，并且能够发现错误并进行纠错，建立或终止节点间的连接，定义节点间的流量控制协议。数据链路的建立、拆除，以及对数据的检错、纠错，是数据链路层的基本任务。

IEEE 802 协议将数据链路层分为两个子层。

- Media Access Control（MAC）层：负责控制网络中的设备如何获得访问权而传输数据。
- Logical Link Control（LLC）层：负责识别网络层协议，进行封装、错误检查、纠错和帧同步。

IEEE 网络（如 802.3 网络、802.11 无线和 802.15.4 ZigBee）的 MAC 子层和 LLC 子层都在数据链路层处理数据。

Point-to-Point Protocol（PPP）协议工作在数据链路层，并且支持多种物理链接。

ITU-T G.hn 规范通过有线连接（电源线、电话线和同轴电缆等）提供了高速的局域网连接，其中就包括完整的数据链路层，负责错误纠正和流量控制等。

### 3. 网络层

网络层将逻辑网络地址翻译成物理主机节点地址，在不同的主机节点间提供各种功能和程序，传输各种不同长度的数据流。网络作为一种介质，多个设备通过这种介质连接在一起，网络中的每个设备都有一个地址，彼此间通过该地址进行连接、信息交互，通过被传递的数据包中含有的目标主机节点地址信息进行准确传送。如果传递的数据太大，在数据链路层不能直接进行传递，那么网络层会将该数据分割成数据段。在传输时，数据段之间没有依赖性，各自独立发送，当到达目标主机节点后进行重组，形成完整的数据。网络层可以发现并报告传输过程中的任何错误，同时提供选项供用户选择是否保证数据可靠传输。

在 Annex 和 ISO 7498/4 国际标准中定义的协议和功能都包含在网络层中，包括路由协议、多播组管理、网络层信息和错误报告、网络地址分配。这些功能都属于网络层，而不是协议。

网络层是可选的，它只适用于多个设备处于不同的、由路由器分隔开的网段，或者当通过应用要求某种网络层或传输层提供的服务、特性或者能力时。

### 4. 传输层

传输层负责跨越多个网络传输不同长度的数据段，并保证服务质量，满足更高层次的要求。例如 Transmission Control Protocol（TCP）协议，可提供可靠的端到端的纠错和流量控制，保证数据的正确传输。

传输层通过链路流量控制、数据分段和组合及错误控制等功能提供可靠性。同时，传输层的一些协议是面向状态或面向连接的，所以可以跟踪数据段的流向，以及在传输失败时重新传输。传输层可以对已经成功传输的数据段进行确认并继续发送下一个数据段，也可以将从应用层收到的消息数据转化成数据包。数据包转化的过程就是将长消息数据分割成短消息数据的过程。

在 OSI 模型中，传输层根据传输协议连接模式的不同，从 TP0 到 TP4 分成 5 类。一般情况下，TP0 代表支持的特性少，不包含错误恢复，多应用于无错误连接；TP4 代表支持的特性多，多应用于可靠性低的网络，例如 Internet 网络，TCP 是 TP4 的典型代表。下表是 TP0～TP4 所支持特性的详细信息。

| 特性名称 | TP0 | TP1 | TP2 | TP3 | TP4 |
|---|---|---|---|---|---|
| 面向连接网络 | 是 | 是 | 是 | 是 | 是 |
| 无线连接网络 | 否 | 否 | 否 | 否 | 是 |
| 串联和分离 | 发 | 是 | 是 | 是 | 是 |
| 分段和重组 | 是 | 是 | 是 | 是 | 是 |
| 错误恢复 | 否 | 是 | 是 | 是 | 是 |
| 重连接 | 否 | 是 | 否 | 是 | 否 |
| 连接复用 | 否 | 否 | 是 | 是 | 是 |
| 数据流控制 | 否 | 否 | 是 | 是 | 是 |
| 超时重传输 | 否 | 否 | 否 | 否 | 是 |
| 可靠传输 | 否 | 是 | 否 | 是 | 是 |

举例来说，传输层和邮局的功能有许多相似之处，如邮局进行邮件分发和分类是根据邮件信封信息来操作的，这与传输层的功能相同。当然，在 OSI 模型中，更高层级可能包含两个"信封"，如加密后的信息，根据第一个"信封"，任何人都可以读取，但是根据第二个"信封"，只有收件人才能读取。一般来说，隧道协议应该发生在传输层，但是根据数据封装地点的不同而不同。如 Generic Routing Encapsulation（GRE）协议，根据功能定义，它好像一个网络层协议；但是它对数据的封装发生在端点，则 GRE 更像一个传输层协议。

### 5．会话层

会话层控制着应用之间的连接，可以启用、管理和终止连接，提供全双工、半双工和单工操作，设置检查点可使通信会话在通信失效恢复时从检查点继续恢复通信，这种功能对于传送大文件极为重要。会话层、表示层和应用层构成 OSI 模型的高三层，面向应用进程提供分布处理、对话管理、信息表示、检查和恢复与语义上下文有关的传送查错等功能。在使用 Remote Procedure Calls（RPC）的应用环境中，会话层经常被用到。

### 6．表示层

表示层的主要功能是将不同格式的信息进行转换，使数据表示具备独立性，成为应用层可以识别的数据。表示层之所以重要，是因为不同体系结构所使用的数据表示方法不同，例如，IBM 主机使用 EBCDIC 编码，而大部分 PC 使用的是 ASCII 码。在这种情况下，就需要表示层来完成这种转换。通过前面的介绍可以总结出，包括会话层在内的下五层完成了端到端的数据传送，并且是可靠的、无差错的传送。但是数据传送只是手段而不是目的，最终是要实现对数据的应用。由于各种应用程序对数据的定义并不完全相同，最容易明白的例子就是键盘，其上的某些按键的含义在许多操作系统中都有差异，表示层就负责消除这些障碍。表示层也可以对数据进行加密。

### 7．应用层

应用层是 OSI 模型中的最高层，最接近终端用户，通过软件应用与终端用户直接通信。其主要功能是识别通信对象、确定资源可用性和同步通信。当通信对象被识别后，应用层确定数据传输应用程序的身份和通信对象的可用性，在确定资源有效性之后，应用层必须确定有足够的网络资源供该请求使用。在同步通信中，应用层管理所有的通信请求。在这个过程中涉及的通信对象、服务的质量要求、用户认证和隐私、语法约束使用等都是由应用程序指定的。

## 3.2　网络介绍

OpenStack 项目中的 Neutron 网络服务组件管理着所有的网络端口，无论是在虚拟网络架构（Virtual Networking Infrastructure，VNI）中，还是物理网络架构（Physical Networking Infrastructure，PNI）中（与 Neutron 连接的部分）。同时，Neutron 网络服务组件可以为租户提供高级服务，如创建防火墙、负载均衡和 VPN（Virtual Private Network）等。

Neutron 网络服务组件利用软件定义提供网络、子网和路由功能，它模仿物理网络设备的功能，在网络中划分多个子网，路由器在不同的子网间传递数据。

路由器通过网关连接网络，虚拟机实例上的网络端口连接到子网中，不同子网中的虚拟机实例通过同一路由器相互连接。

已经部署完成的 Neutron 网络服务组件至少包含一个外部网络（External Network），外部网络不仅是一个网络，它同时代表着一个物理网络的视图。外部网络能够连接 OpenStack 项目之外的网络环境，DHCP 功能被关闭。

除了外部网络外，还至少包含一个内部网络，虚拟机实例在这些内部网络中获得 IP 地址，那些连接在相同路由器上的设备能够与虚拟机实例相互通信。

外部网络和虚拟机实例之间的通信需要由路由器连接，路由器依靠网关完成。例如一个物理路由器设备，不同子网中的设备之间进行通信必须依靠路由器，虚拟机实例能够访问外部网络必须通过网关完成。

在外部网络中，也可以分配 IP 地址给虚拟机实例，虚拟机实例的网络端口既可以连接内部网络，又可以连接外部网络。这样，网络中的所有设备都可以与该虚拟机实例进行通信。

Neutron 网络服务组件支持安全组（Security Group）设置，管理员可以定义一些安全规则，包括禁止或者允许某个端口、端口范围和访问类型（如 TCP 或 UDP）。多个安全组可以应用于同一个虚拟机实例。

Neutron 网络服务组件支持的每个 plug-in 都有各自的概念、功能和作用，可能不会全部应用于 OpenStack 项目，但是理解和掌握这些 plug-ins 对 Neutron 网络服务组件的应用会起到事半功倍的作用。Neutron 网络服务组件包含一个核心的网络 plug-in 和一个安全组 plug-in，Firewall as a Service（FWaaS）和 Load Balancer as a Service（LBaaS）plug-ins 都可以作为 Security Group plug-in 的一个选项。

## 3.3 网络架构

在部署和使用 Neutron 网络服务组件前，理解 Neutron 网络服务组件的架构和拓扑，以及与 OpenStack 项目中的其他服务组件如何交互和互动，将对后期的学习、工作很有帮助。

Neutron 网络服务组件是 OpenStack 项目中一个独立、核心的模块化服务组件，它与其他服务组件一起协同工作，如 Nova 计算、Glance 镜像、Keystone 身份认证和 Horizon 仪表板服务组件等。它可以在多个物理主机节点上部署多个服务。在网络主机节点上运行 neutron-server 主进程提供了相关程序需调用的 Networking API 和进行配置网络插件的管理。

在控制节点上可以运行 Nova 计算服务组件，同样，也可以将 Neutron 网络服务组件部署在控制节点上。当然，Neutron 网络服务组件是独立组件，支持运行在专用的物理主机节点上，而且最好的实践也是部署在专用的物理主机节点上。Neutron 网络服务组件包括下表所示的 agents。

| agent | 描 述 |
| --- | --- |
| plug-in agent (neutron-*-agent) | 运行在每个计算节点上，按照本地虚拟交换机的配置执行操作。插件不同，agent 也可能不同。一些插件不需要 agent |
| dhcp agent (neutron-dhcp-agent) | 为租户网络提供 DHCP 服务 |
| L3 agent (neutron-l3-agent) | 提供 L3/NAT 转发，使租户内的虚拟机实例被外部网络访问 |
| metering agent (neutron-metering-agent) | 提供 L3 数据流量的监控、计量 |

以上这些 agents 通过 RPC（RabbitMQ 或 Qpid）或标准 API 与 Neutron 网络进程进行通信，在运行过程中，需要与其他服务组件进行交互。

- Neutron 网络服务组件依赖于 Keystone 身份认证服务组件的功能，对所有 API 请求进行身份认证和授权。
- Nova 计算服务组件通过调用标准 API 与 Neutron 网络服务组件进行交互。在创建虚拟机实例时，nova-compute 服务与 Neutron 网络服务组件的 API 进行交互，以分配虚拟网络地址给虚拟机实例，用于连接到网络中。
- Horizon 仪表板服务组件与 Neutron 网络服务组件的 API 进行交互，管理员和租户使用基于 Web 的图形界面管理网络服务。

## 3.4 网络 API 简介

Neutron 网络服务组件是一个虚拟网络服务，它提供了一组强大的 API 来定义网络连接和 IP 地址，供 Nova 计算服务组件使用。

Nova 计算服务组件也提供了一组强大的 API 来描述虚拟机的计算资源。同样，Neutron 网络服务组件提供的 API 通过网络、子网和网络端口等抽象概念而详细、具化地描述网络资源，如下表所示。

| 资　源 | 说　明 |
| --- | --- |
| 网络 | 基于 OSI 模型二层独立的区段，类似于物理网络交换机中的 VLAN |
| 子网 | 一组基于 IPv4 和 IPv6 的地址段及它们的相关配置状态 |
| 网络端口 | 连接设备的连接点，例如连接到虚拟网络中虚拟机的网络端口，以及与该端口相关的网络配置，包括 MAC 地址和 IP 地址 |

在配置网络拓扑结构时，可以创建网络、子网和网络端口，以配合其他服务组件使用。例如，Nova 计算服务组件可以将虚拟设备连接到虚拟网络端口，使其可以连接到虚拟网络中。

Neutron 网络服务组件支持每个租户有属于自己的网络地址段，每个租户可以根据自己的需求选择合适的网络地址类型和区段。每个租户之间的网络设置可以重复，使用同一类型或区段的 IP 地址。

Neutron 网络服务组件的功能可以归纳为以下三点：

- 具备高级的虚拟网络功能。例如，构建多层次的 Web 应用、应用系统从传统环境迁移到云环境无须修改 IP 地址等。
- 灵活性高，在网络架构、功能方面提供更多自定义。
- 允许开发者对 Neutron 网络服务组件进行二次开发，从而使更多人受益。

## 3.5　LBaaS 和 FWaaS

Load Balancer as a Service（LBaaS）功能使 Neutron 网络服务组件在多个网络主机节点间可以均衡分配、处理请求，这种机制可以分散负载压力，使请求造成的工作负载被均衡分配，以提高系统资源的利用率。负载均衡包含以下三种策略。

- Round Robin：在多个网络主机节点间轮询处理各种请求。
- Source IP：来自同一源 IP 的请求被持续指向同一网络主机节点。
- Least Connections：分配请求给具有最少活动连接数的网络主机节点。

LBaaS 的特性描述如下表所示。

| 特 性 | 描 述 |
| --- | --- |
| 监控 | LBaaS 通过 ping、TCP、HTTP 和 HTTPS 等方法提供有效的监控，以确保各成员是否可以有效地处理请求 |
| 管理 | LBaaS 使用各种工具进行管理。用户可以通过 CLI（Neutron）和仪表板（Horizon）进行负载均衡的管理操作 |
| 连接限制 | 请求入口信息流可以使用连接限制进行控制。该功能可以控制工作负载压力，同时也可以减轻 DoS（Denial of Service）攻击 |
| Session 持久性 | LBaaS 通过路由请求到相同网络主机节点，从而保证 Session 具有持久性和有效性。其支持基于 Cookie 和源 IP 地址的路由方法 |

Firewall as a Service（FWaaS）添加防火墙管理到 Neutron 网络服务组件中，使用 Iptables 作为防火墙策略应用到 L3 虚拟路由中。

安全组（Security Group）应用于虚拟机实例级别，FWaaS 在虚拟路由外围过滤各种流量。

下图为 instance02 虚拟机实例进口和出口流量的数据流图。

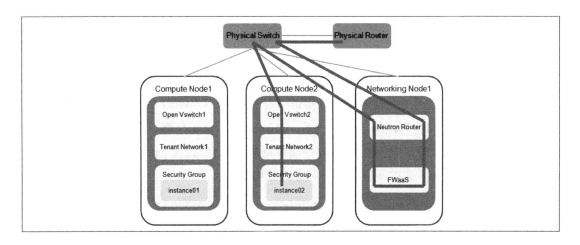

## 3.6 网络类型介绍

OpenStack 项目中的网络分为两类，分别为 Tenant Network 和 Provider Network。普通用户可以在租户内创建 Tenant Network，并且关于该 Tenant Network 的信息对其他用户是不可见的，以保证独立性和安全性。具有管理员权限的用户可以创建 Provider Network，根据已经存在的物理网络属性设置该 Provider Network 的属性和信息，该虚拟网络必须与数据中心已经存在的物理网络配置相匹配。具体的拓扑结构如下图所示。

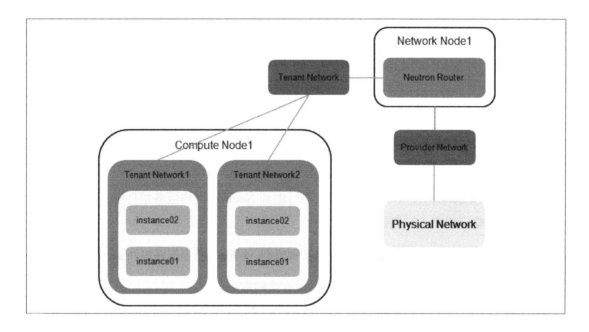

### 1．Provider Network

Provider Network 允许具有管理员权限的用户创建虚拟网络直接连接到数据中心的物理网络中，虚拟机实例利用此网络可以直接访问外部网络。Provider Network 扩展属性允许管理员管理虚拟网络和物理网络之间的关系，如 VLAN、Tunnels 等。在启用这些扩展属性后，具有管理员权限的用户便可以使用这些属性创建虚拟网络。Open vSwitch（OVS）和 Linux Bridge 插件都支持该扩展属性，且配置大概相同。

Provider Network 支持的网络类型包括 Flat（untagged）和 VLAN（802.1Q tagged）。

## 2．Tenant Network

普通用户可以在租户内创建虚拟网络，在默认情况下，租户之间的网络是隔离的、不能共享的。

Neutron 网络服务组件支持的网络隔离和重叠类型（overlay）如下。

- Flat：所有虚拟机实例都连接在同一网络中，并且与宿主机也可以运行在同一网络中，没有对网络数据包进行打标签（vlan tag）或者隔离。

- Local：虚拟网络使用 Nova 计算服务组件的 nova-network。目前该网络类型的代码不再更新，功能使用很少。

- VLAN：Neutron 网络服务组件使用 vlan tag（802.1Q tagged）技术允许用户创建多个 Provider Network 和 Tenant Network，该虚拟网络中的 vlan tag 信息与数据中心物理网络中的 vlan tag 信息是一一对应的。因此，虚拟机实例可以和网络中的任何设备进行通信，包括服务器、防火墙、负载均衡和其他 L2 层的设备。

- GRE 和 VXLAN：GRE 和 VXLAN 是一种封装数据包的协议，创建重叠网络以激活和控制虚拟网络。Neutron 网络服务组件中的 neutron router 允许采用 GRE 和 VXLAN 的 Tenant Network 数据流出虚拟网络或当下租户网络，使数据自由流动，原因就是 neutron router 连接了 Tenant Network 和外部网络，使彼此互通。从外部网络访问 Tenant Network 中的虚拟机实例，需要通过 Floating IP 地址进行连接。

下表对本章中出现的网络名词进行了总结。

| 名　　词 | 描　　述 |
| --- | --- |
| Virtual Network（虚拟网络） | 基于 L2 层网络（由 UUID 唯一标识，名字是可选的），运行在其中的虚拟网络端口可以分配给虚拟机实例和各 agents 使用。Open vSwitch（OVS）和 Linux Bridge（桥接）都支持多种配置机制实现虚拟网络 |
| Physical Network（物理网络） | 物理主机节点和其他网络设备相互连接的网络。每个物理网络都可以支持多个虚拟网络。Provider Network 扩展属性和插件配置使用字符串标识物理网络 |
| Tenant Network（租户网络） | 由普通用户或管理员用户创建的虚拟网络。租户之间的网络是隔离的、不共享的 |

续表

| 名词 | 描述 |
| --- | --- |
| Provider Network（供应商网络） | 由具有管理员权限的用户创建的虚拟网络，但是该虚拟网络的属性信息与数据中心内的物理网络属性信息必须相同。通常被用作直接访问非 OpenStack 管理的设备 |

ML2、Open vSwitch（OVS）和 Linux Bridge（桥接）插件支持 VLAN、Flat 和 Local 三种虚拟网络类型。只有 ML2 和 Open vSwitch（OVS）支持 GRE 和 VXLAN 虚拟网络类型，支持的前提是主机节点 Kernel、Open vSwitch（OVS）和 iproute2 包含其所需的特性。

# 第 4 章
# 存储服务介绍

顾名思义，存储服务是为数据保存而使用的，在 OpenStack 项目中被多个服务组件使用，非常重要。从数据保存时间的角度区分，存储可以分为两种：临时存储和持久存储。

临时存储是指数据被虚拟机实例使用，一旦虚拟机实例被关机、重启或删除，该虚拟机实例中的所有数据信息全部丢失。换言之，就是所有的数据都保存在临时存储中。在 OpenStack 项目中，部署完 Nova 计算服务组件后，用户可以使用 nova boot 命令创建虚拟机实例，这时虚拟机实例使用的就是临时存储，所有数据都保存在临时存储上，安全性没有保障。

持久存储包括对象存储、块存储和文件系统存储，它们维护数据持续可用，保证数据安全性，无论虚拟机实例是否终止。

下表从多个角度对存储特性进行了描述。

| 特 性 | 临时存储 | 块存储 | 对象存储 | 文件系统存储 |
| --- | --- | --- | --- | --- |
| 用途 | 运行操作系统或暂存数据空间 | 作为持久性存储提供给虚拟机实例使用 | 存储数据，包括虚拟机镜像 | 作为持久性存储提供给虚拟机实例使用 |
| 访问方式 | 文件系统 | 块设备，能够进行分区、格式化和被挂载等 | API 调用 | 共享的文件系统服务，能够进行分区、格式化和被挂载等 |

续表

| 特 性 | 临时存储 | 块存储 | 对象存储 | 文件系统存储 |
|---|---|---|---|---|
| 访问途径 | 虚拟机实例中 | 虚拟机实例中 | 任何地方 | 虚拟机实例中 |
| 管理者 | Nova 计算服务组件 | Cinder 块存储服务组件 | Swift 对象存储服务组件 | Manila 共享文件系统存储 |
| 数据持久性 | 虚拟机实例被关机 | 删除 | 删除 | 删除 |
| 容量大小确定 | 根据管理员对主机类型（flavor）的定义 | 根据需求定义 | 所有有效物理存储容量 | • 根据需求定义<br>• 扩大需求<br>• 用户磁盘配额<br>• 管理员设置磁盘容量限制 |
| 加密方式 | nova.conf 文件中参数配置 | 管理员启用卷加密功能 | 不支持 | 不支持 |
| 使用案例 | 小容量磁盘 | 大容量磁盘 | 超大容量存储 | 依赖于后端存储容量，当开启 thin provisioning 功能后，可以提供部分容量（不是全部申请容量） |

用户对 OpenStack 环境中存储的使用存在许多不同的需求，一些用户需要快速访问非经常修改的数据，一些用户需要访问可以被格式化成文件系统的存储，还有一些用户希望虚拟机实例不再使用时存储能够立即释放。针对目前主流的开源存储技术特点，我们对其进行了总结，如下表所示。

| 存储技术 | 对象（Object） | 块设备（Block） | 文件（File-Level） |
|---|---|---|---|
| Swift | ✓ | | |
| LVM | | ✓ | |
| Ceph | ✓ | ✓ | |
| Gluster | ✓ | ✓ | ✓ |
| NFS | | ✓ | ✓ |
| ZFS | | ✓ | |
| Sheepdog | ✓ | ✓ | |

## 4.1 块存储（Cinder）服务介绍

OpenStack 项目中的 Cinder 块存储服务组件为虚拟机实例提供了块存储设备，同时为管理块存储设备提供了一整套方法，如卷快照（Volume Snapshot）、卷类型（Volume Type）等。块存储类型是由驱动或者后端设备配置的驱动决定的，如 NAS、NFS、SAN、ISCSI、Ceph 或其他。Cinder 块存储服务组件的 API 和 cinder-scheduler 服务通常运行在控制节点上，cinder-volume 服务可以运行在控制节点、计算节点或者独立的存储节点上。

Cinder 块存储服务组件由以下服务进程组成。

- cinder-api：接收 API 请求，并将请求转发到 cinder-volume。
- cinder-volume：与块存储直接进行交互通信，处理一些任务，如由 cinder-scheduler 分配的任务，同时通过消息队列与这些任务进行交互通信。cinder-volume 还会维护块存储的状态，通过驱动与各种类型的存储进行交互通信。
- cinder-scheduler daemon：选择最佳存储节点创建卷。具有相似功能的是 nova-scheduler。
- cinder-backup daemon：提供任意类型卷的备份。
- messaging queue：负责在各进程间传递信息。

块存储又称卷存储（Volume Storage），为用户提供基于数据块的存储设备访问，用户对数据块存储设备的访问和交互是通过将数据块存储设备映射到正在运行的虚拟机实例上进行的，可以对其进行读/写、格式化等。

块存储属于持久存储，当取消块存储设备与虚拟机实例之间的映射或将块存储设备重新映射给其他虚拟机实例时，该块存储设备上的数据都不受影响，数据是完整的。换言之，只要该块存储设备是存在的、无损坏的、完整的，那么其上的数据就是完整的、可用的。至于该块存储设备是否映射给其他虚拟机实例识别和使用，与其上已经保存的数据的完整性和可用性无关。块存储由 OpenStack 项目中的 Cinder 块存储服务组件提供，根据已经包含的存储驱动，其目前支持多种类型的后端存储。

许多存储驱动支持虚拟机实例直接访问底层存储，无须经过层层转化和性能消耗，这无疑对提升整体的 I/O 性能很有帮助。同时，Cinder 块存储服务组件也支持使用文件作为块设备，如 NFS、GlusterFS 等。在 NFS 和 GlusterFS 文件系统中，可以创建一个独立文件作为块设备映射给虚拟机实例使用。这种方法类似于在 QEMU 中创建虚拟机实例，这些虚拟机实例是一个个文件，保存在/var/lib/nova/instances 目录中。

## 4.2 对象存储（Swift）服务介绍

OpenStack 项目中的 Swift 对象存储服务组件通过 REST API 提供对象数据存储和检索。它不能独立使用，至少要和 Keystone 身份认证服务组件配合使用，所以在部署 Swift 对象存储服务组件前，Keystone 身份认证服务组件必须已经准备就绪。

Swift 对象存储服务组件支持多租户（Multi-Tenant），投入成本低，具有高扩展性和存储大量非结构化数据的特性。

Swift 对象存储服务组件包括以下几部分。

- Proxy servers(swift-proxy-server)：接收对象存储的 API 和 HTTP 请求，修改 metadata、创建 containers；在 Web 图形界面上提供文件或者 containers 列表；使用 MemCached 提供缓存功能，提高性能。
- Account servers（swift-account-server）：管理对象存储内的账户。
- Container servers（swift-container-server）：管理对象存储内 containers 和文件夹之间的映射。
- Object servers（swift-object-server）：管理真正的对象数据，如文件等。
- Various periodic processes：执行日常事务。其中的复制服务能够保证数据的连续性和有效性，还包括审核服务、更新服务和删除服务。
- WSGI middleware：处理认证相关问题，与 Keystone 身份认证服务组件连接。
- swift client：允许各种具有权限的用户在该客户端上提交命令，执行操作。
- swift-init：初始化 Ring 文件的脚本，需要守护进程名字作参数，并提供操作命令。
- swift-recon：CLI 工具，用于检索集群各种性能指标和状态信息。

- swift-ring-builder：创建和重平衡 Ring 的工具。

在 Swift 对象存储服务组件中，用户通过调用 API 可以访问存储中的元数据，与著名的 Amazon S3 对象存储有异曲同工之处。Swift 对象存储的主要功能是对静态大数据集的归档和管理，同时也可以存储虚拟机镜像。

Swift 对象存储服务组件提供了一套可扩展、高可用的存储方案，以代替传统的、昂贵的存储设备。从本质上讲，此类型存储的诞生为解决不同严重级别和时间点的硬件故障提供了另外一种方案，如控制器宕机、RAID 卡损坏等。

Swift 对象存储具备数据持久性和可用性的特点，依靠软件逻辑设计使所有数据能够均衡分布存储。一般情况下，默认保存三份数据。这三份数据的保存位置非常重要，既可以分布在同一服务器的不同硬盘中，也可以分布在同一机架内的不同服务器中，分布位置的选择对集群整体性能的影响很大。在 Swift 对象存储集群中，当出现存储数据的主机节点宕机时，整个集群的负载压力会非常大。例如，在一个含有三份副本的集群中，一个主机节点上保存有 8TB 数据，那么在故障出现时，瞬间便会有 24TB 的数据需要进行传输和重平衡，可想而知，对网络和磁盘的压力有多大。这就对网络带宽和磁盘性能提出了更高的要求，在生产环境中要尽可能多地使用网卡聚合和固态磁盘等技术来提高整体性能。

## 4.2.1 对象存储特点

Swift 对象存储区别于传统存储（见下图），具体如下：

- 所有的数据存放和访问都有一个 URL 地址。
- 所有的数据保存三份副本，分布在不同的 Zone 里。
- 所有的数据都有自己的元数据（metadata）。
- 软件开发人员可以通过 API 与 Swift 对象存储进行交互操作。
- 数据随机存放在集群中的任意位置。
- 集群添加存储节点不产生性能损耗，是横向扩展，而不是纵向叠加。
- 数据不需要一次性迁移到新存储。

- 集群添加存储节点不需要停机时间。
- 故障存储节点或硬件设备进行替换不需要停机时间。
- Swift 对象存储运行在标准的 x86 服务器上,如 Dell、HP 和 Lenovo 服务器。

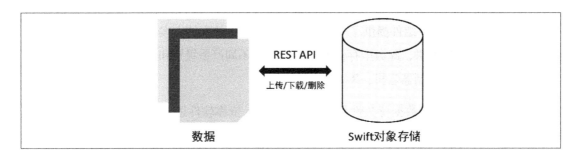

软件开发人员既可以通过 API 也可以通过开发语言的函数库,与 Swift 对象存储进行交互操作,开发语言包括 Java、Python、Ruby 和 C#等。Amazon S3 和 RackSpace 的对象存储与 Swift 对象存储非常相似。

下表是 Swift 对象存储的特点和优势。

| 特　　点 | 优　　势 |
| --- | --- |
| 与商业存储方案对比 | 低成本,避免厂商锁定 |
| 硬件或节点故障自动保护 | 自我恢复,可靠性高,数据冗余保护 |
| 无限容量 | 具备高扩展性和大而扁平的命名空间 |
| 多个维度扩展 | 向外扩展的机构,垂直和水平分布式存储,备份和归档大量数据 |
| 全新架构(账户/容器/对象) | 非树形结构存储,可以扩展到 PB 或更大容量级别 |
| 与传统数据冗余方案对比 | 软件逻辑设计,数据副本数量可以配置、修改 |
| 增加容量便捷 | 弹性增加,无停机时间 |
| 没有控制单点 | 高性能,没有瓶颈 |
| 不要求 RAID 卡 | 处理大量小而随机的读和写 |
| 内在管理工具 | 创建、修改和删除用户,上传和下载容器,对容量、网络、主机节点和集群健康进行监控 |
| 数据有效期限 | 对数据的访问,用户可以设置有效期或一个 TTL |

续表

| 特　点 | 优　势 |
| --- | --- |
| 直接访问数据 | 通过 URL 可以直接访问数据 |
| 实时性 | 实时查看和接收用户的请求 |
| S3 API 兼容性 | 利用工具兼容 S3 API |
| 管理每个账户的容器 | 控制对容器的使用和访问 |
| 对 NetApp、Nexenta 和 Solidfire 的支持 | 统一支持全部块设备存储 |
| 块设备的备份和快照 | 数据保护和数据恢复 |
| 独立性 | 分离访问终端和 API 调用 |
| 与 Nova 计算服务组件集成 | 完全支持 Nova 计算服务组件，连接块设备，生成使用率报告 |

## 4.2.2　对象存储组成

Swift 对象存储使用以下组件提供高可用性、高稳定性和高并发。

- Proxy Server：接收和处理各种传入的 API 请求。

- Ring：将数据名称映射到磁盘的具体位置上。

- Zone：用于隔离副本数据，当一个 Zone 出现问题时，不会影响其他 Zone 中的副本数据。

- Account 和 Container：每个 Account 和 Container 都是一个独立的数据库，分布在集群的多个存储节点上。每个 Account 数据库包含属于它的 Container 列表，每个 Container 数据库包含属于它的对象数据列表。

- Object：数据。

- Partition：每个 Partition 中保存对象数据、Account 数据库和 Container 数据库，管理着对象数据在集群中的存放位置。

下图是对象存储组成图示及操作层级展示。

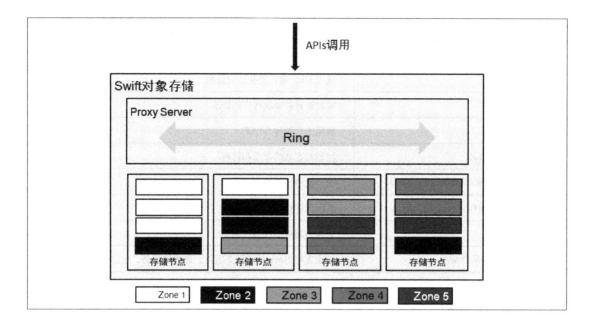

### 1．Proxy Server

Proxy Server 是 Swift 对象存储的公共接口，接收并处理所有传入的请求。当 Proxy Server 接收到请求后，根据对象数据的 URL 定位到相应的存储节点上，如 https://swift.object.storage/v1/account/container/object。同时，Proxy Server 负责协助处理请求反馈、请求失败和标记时间戳等。

Proxy Server 采用共享无状态架构，能够根据业务负载压力扩展 Proxy Server，最小的 Proxy Server 集群包含两个主机节点，当其中一个主机节点出现故障时，另一个主机节点能够接管所有业务请求。

### 2．Ring

Ring 是对象数据的名称与其在磁盘上具体位置之间的映射。因为 Account、Container 和 Object 之间的组合会产生许多 Rings，当需要对 Account、Container 或 Object 进行操作时，都需要与相应的 Ring 进行交互操作，以确定其在磁盘上的具体位置。

Ring 使用 Zone、Partition、磁盘和数据副本构成映射，Ring 中的 Partition 都有多个副本，默认是三份副本。当出现故障切换时，Ring 可以确定使用哪个磁盘进行数据迁移。

Ring 使用 Zone 对数据进行隔离，每个 Partition 的副本都存储在不同的 Zone 中。Zone 可以是一块磁盘、一台服务器，甚至是一个数据中心。

Ring 将数据副本的 Partition 存放在多个磁盘上，当 Partition 进行移动时（如集群添加新存储节点），Ring 需要确认移动的 Partition 数量，一次只允许移动一个副本的 Partition。通过设置权重值也可以平衡 Partition 的分布，控制数据流向，避免集群在压力负载大时各主机节点的压力负载不均衡。下图是 Ring 的功能示意图。

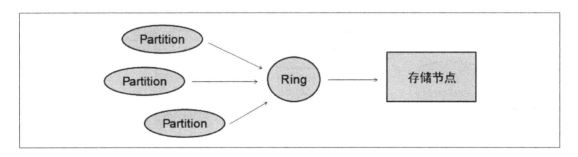

### 3. Zone

Swift 对象存储利用 Zone 功能隔离风险，当某一 Zone 内的数据丢失或被破坏时，不影响其他 Zone 内的数据副本，保持数据整体上的完整性和可用性。理想情况下，每个数据副本分别存储在不同的 Zone 中，充分隔离。Zone 的最小单位可以是一块磁盘设备或一组磁盘设备，其标志性作用是允许存储节点出现停机时间而数据仍然保持实时可用性。下图是 Zone 的功能示意图。

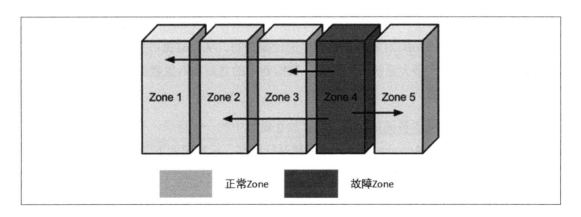

### 4．Account和Container

Account 和 Container 都有自己独立的 SQLite 数据库，SQLite 数据库采用分布式，部署在 Swift 对象存储集群的多个存储节点上。每个 Account 数据库中包含属于它的 Container 列表，同样，每个 Container 数据库中也包含属于它的对象数据列表，示意图如下图所示。

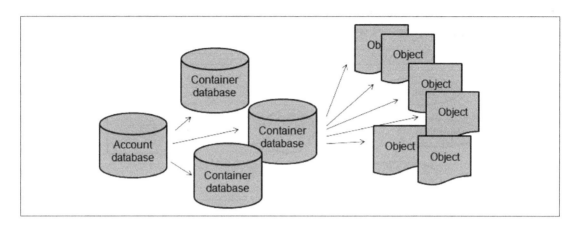

由上图可以看出，为了能够正确获取对象数据位置，Account、Container 和 Object 之间的对应关系很重要。

### 5. Partition

Partition 代表需存储数据的一个集合，包含多个 Account 数据库、Container 数据库和相应的对象数据。

可以把 Partition 看作一个小型仓库，包含多个 Account、Container 和相应的对象数据，它是作为一个整体在 Swift 对象存储集群中进行操作的，从而避免单独操作多个小型的对象，既增加复杂度，又增大系统负载压力。数据副本的复制和对象数据的上传下载都是基于 Partition 进行操作的。

关于 Partition 的概念很简单，它就像存放在磁盘中的一个目录，有自己的哈希表，关联着属于该 Partition 的 Account、Container 和对象数据，示意图如下图所示。

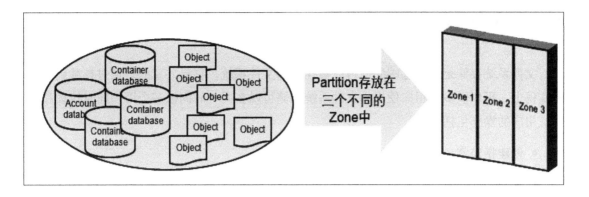

## 6．Replicator

为保证存放在 Swift 对象存储集群中三份数据副本的一致性，Replicator 会持续检查每个 Partition，将本地的 Partition 与其他 Zone 中的 Partition 副本进行比较，检查是否一致。

Replicator 通过检查哈希表确认是否需要进行数据重同步，是否发起数据同步操作。每个 Partition 都有一张哈希表，包含一个含有哈希表的目录。例如，数据存储三个副本，对于其中一个 Partition，会将其哈希文件与其他两个副本的哈希文件进行比较，如果不同，则数据需要进行重同步，目录中的哈希文件也需要进行同步。

Replicator 会持续检查数据的一致性，其中，新数据的同步优先级较高。

当集群中某个存储节点宕机后，包含同样数据副本的存储节点会进行检查和发出通知，并将数据复制到其他存储节点上。示意图如下图所示。

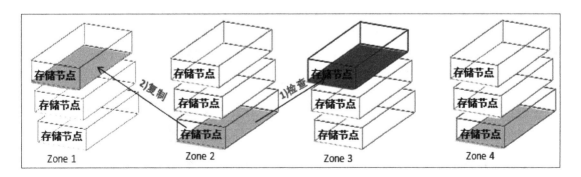

## 4.3 文件系统存储

文件系统存储是一个远端的、可以被挂载的文件系统。它是共享的，通过挂载到虚拟机实例上，可以供多个租户使用。文件系统存储可以在同一时间被多个用户同时挂载和访问。用户的操作包括：

- 创建指定容量大小的文件和文件系统协议。
- 创建的文件可以分布在一个或多个服务器上。
- 指定访问规则和安全协议。
- 支持快照。
- 通过快照恢复一个文件系统。
- 查看使用率。

文件系统存储与 Cinder 块存储相同，也属于持久存储。它能被多个虚拟机实例挂载，可以从虚拟机实例解除映射，重新映射到其他虚拟机实例，而数据依然保持完整。

在 OpenStack 项目中，文件系统存储的程序代号是 manila，支持多种后端存储驱动，通过多种存储协议（NFS、CIFS、GlusterFS 和 HDFS）进行共享。

## 4.4 Ceph 简介

Ceph 是一个符合 POSIX (Portable Operating System for UNIX®)、开源的分布式存储系统，遵循 LGPL 协议。该项目最初由 Sage Weill 于 2007 年开发，其理念是开发一个没有任何单点故障的集群，确保能够跨集群节点进行永久数据复制。

与任何经典的分布式文件系统中一样，放入集群中的文件是条带化的，依据一种称为 Ceph Controlled Replication Under Scalable Hashing（CRUSH）的数据分布算法放入集群节点中。

Ceph 支持自动扩展、自动恢复和自主管理集群，其生态系统支持许多与其交互的方式，这使得在已运行的基础架构中进行集成变得既简单又便捷，即便它执行的是一个在项目中需要提供块和对象双类型存储功能的复杂任务。Ceph 共有 5 个主要组成部分，如下图所示。

- RADOS（Reliable Autonomic Distributed Object Store）是一个稳定的、独立的和完全分布式的对象存储，具备自主健康检查、自主恢复、自主管理和高级智能等特点。因此，理解 RADOS 是理解 Ceph 的基础与关键。
- LIBRADOS 是一个 lib 函数库，运行应用程序直接访问 RADOS，支持 PHP、Ruby、Java、Python、C 和 C++ 等编程语言与其交互。
- RADOSGW 提供了兼容 Amazon S3 和 Swift 的 RESTful API 接口，相应的对象存储可与其交互通信，将对象数据直接存储在集群中。
- RBD 是一个稳定的、完全分布式的块设备，包含 Linux 内核客户端和 QEMU/KVM 驱动，为物理主机或虚拟机提供块存储。
- Ceph FS 是一个与 POSIX 兼容的分布式文件系统，包含 Linux 内核客户端和 FUSE（Filesystem in Userspace）支持。

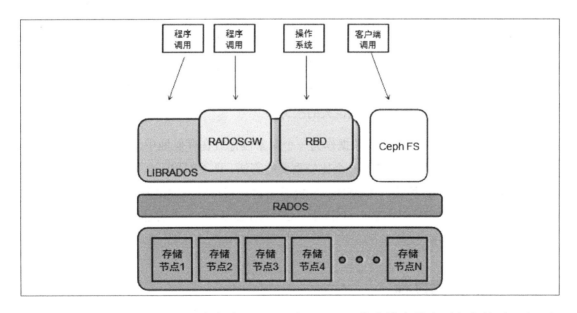

当我们在 OpenStack 云平台上使用 Ceph 或 Ceph FS 作为块存储和对象存储时，需要安装部署多个 Ceph 存储节点、配置网络和 Ceph 存储集群。一个 Ceph 存储集群要求至少包

含一个 Ceph Monitor 进程和两个 Ceph OSD 进程。当使用 Ceph FS 时，还需要使用 Ceph 元数据服务器（MDS），如下图所示。

```
┌─────────────┐  ┌─────────────┐  ┌─────────────┐
│    OSDs     │  │   Monitor   │  │     MDS     │
└─────────────┘  └─────────────┘  └─────────────┘
```

下面分别介绍 OSD、Monitor 和 MDS。

- OSD：一个 Ceph OSD 进程负责存储数据，处理数据复制、恢复、回填、重平衡等，同时，通过心跳检查 Ceph OSD 进程的相关状态信息并汇总传递给 Ceph Monitor 进程。当 Ceph 存储集群设置数据存储副本数量为 2 时（Ceph 存储集群默认存储三份数据副本，但是可以根据需要进行相应调整），那么该 Ceph 存储集群要求至少运行两个 OSD 进程，以达到 Active+Clean 的状态。

- Monitor：Ceph Monitor 维护着关于集群状态的映射管理表（map），包括 Monitor map、OSD map、PG（Placement Group）map 和 CRUSH map。同时，在 Ceph Monitor、Ceph OSD 和 PG 中，Ceph 还维护着每次状态更改的历史信息（epoch）。

- MDS：Ceph 元数据服务器存储着 Ceph FS 需要使用的元数据（Ceph 块存储和对象存储不需要使用 MDS），POSIX 文件系统用户可以在 Ceph FS 上执行 ls、find 等命令，这一切都要归功于 Ceph 元数据服务器，是它帮助用户实现了这种功能，同时又避免了对 Ceph 存储集群造成巨大的压力。

在 Ceph 存储集群中，用户的数据作为一个对象被存放在存储池中，至于数据存放在什么位置、以何种方式存放，就涉及一个比较重要的算法，即 CRUSH（Controlled Replication Under Scalable Hashing）算法。CRUSH 算法能够最优地安排对象数据放置在某个存放组中，并进一步安排该存放组放置在某个 OSD 中。CRUSH 算法使 Ceph 存储集群具备了高扩展、重平衡和动态恢复的能力。

## 4.4.1 存储数据过程

在 Ceph 存储集群中，数据的来源包括 Ceph 块存储（RBD）单元组件、Ceph 对象存储单元组件、Ceph FS 和使用 librados 开发的应用程序，数据作为一个对象进行存储。每个对象关联着一个文件，该文件存储在磁盘中，Ceph 的 OSD 进程负责处理磁盘中对象的读/写操作，如下图所示。

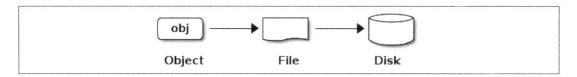

Ceph OSD 进程以扁平命名空间格式（而不是传统的目录层次结构）存储着所有的对象数据，一个对象数据拥有一个唯一 ID、二进制数据和元数据（以 name-value 键值对形式出现），如下图所示。Ceph FS 使用元数据存储文件属性，文件属性包括文件属主、创建日期、修改日期等。

| ID | Binary Data | Metadata | |
|---|---|---|---|
| 1234 | 010101010101010011010101010010<br>010110000101010011010101010010<br>010110000101010011010101010010 | name1<br>name2<br>nameN | value1<br>value2<br>valueN |

### 4.4.2 可扩展性和高可用性

对于传统的分布式存储，客户端通过一个中央控制器与其进行通信，中央控制器是一个单点输入，控制着整个存储集群的功能。这就暴露出其在扩展性和性能方面的一个缺陷，即中央控制器作为一个单点，一旦出现宕机或其他意外情况，整个集群就会出现功能性宕机，不能对外提供服务。

为了解决此问题，Ceph 使客户端可以直接与 OSD 进程通信，OSD 进程存在于集群中的每个存储节点中，负责创建对象数据副本，以此保证数据的安全性和高可用性。同时，集群中的 Monitor 进程负责整个集群的高可用性，使用 CRUSH 算法确保 Ceph 存储集群具备中央控制器所提供的功能。

#### 1．CRUSH 介绍

Ceph 的客户端和 OSD 进程都需要使用 CRUSH 算法计算出对象数据的存放位置，而不是像传统的分布式存储一样，完全依赖于中央控制器。同时，CRUSH 算法提供了一个更好的数据管理方式，即分配工作负载给 Ceph 集群中所有的 OSD 进程，以实现更大规模的集群。Ceph 集群中的副本技术也是由 CRUSH 算法提供的，数据保存多个副本保证了安全性和可扩展性。

### 2．集群映射表

Ceph 依赖于客户端和 OSD 进程来实现集群的拓扑，其包括 5 个 map。

（1）Monitor map：其包含了 fsid、位置、名称地址、端口号、当前的 epoch、map 创建日期和最后修改日期。可以使用"ceph mon dump"命令查看 Monitor map。

（2）OSD map：其包含了 fsid、map 创建日期和最后修改日期、存储池、副本数量、OSD 列表及其数量。可以使用"ceph osd dump"命令查看 OSD map。

（3）PG map：其包含了 PG 版本、时间戳、最后的 OSD map epoch 和每个 PG 的详细信息等。PG 的详细信息包括 PG ID、状态和使用率统计等。

（4）CRUSH map：其包含了存储设备列表、故障域（如设备级别、主机级别、机柜级别和数据中心级别等）和用于存储数据时映射层级传递的规则。可以使用"ceph osd getcrushmap -o {filename}"命令查看 CRUSH map。

（5）MDS map：其包含了 MDS map epoch、map 创建日期和最后修改日期、元数据池、元数据服务列表及状态。可以使用"ceph mds dump"命令查看 MDS map。

每个 map 中都包含着操作状态更改的历史信息。Ceph Monitor 维护着集群映射表的主副本，其包括集群成员、状态、修改和存储集群的健康状态。

### 3．Monitor高可用

在 Ceph 客户端读/写数据之前，必须首先与 Monitor 进程建立联系，以获取最新的集群映射表副本。Ceph 存储集群可以只包含一个 Monitor 进程，但是这样又出现了与传统分布式存储同样的问题，即单点故障。一旦该 Monitor 进程被终止，Ceph 客户端将读/写数据失败。

为了解决此问题，增加 Monitor 稳定性和容错性，Ceph 存储集群支持 Monitor 集群功能。但是随着 Monitor 集群规模和负载压力的增大，在 Monitor 集群中，延迟和其他故障可能导致 Monitor 集群状态出现异常，进而导致整个 Ceph 存储集群出现故障。针对该问题，Ceph 设置了一个协议，采用少数服从多数规则和 Paxos 算法以保证 Monitor 集群的可持续性，进而不影响整个集群的状态，使 Ceph 存储集群可以持续对外提供服务。

### 4．身份认证

为了进行用户身份认证，同时避免存储集群遭受人为攻击，Ceph 提供了 cephx 认证系统，以验证用户和进程身份是否合法。cephx 使用共享的安全 keys 进行身份认证，客户端和 Monitor 集群各自拥有同一个安全 key，客户端和 Monitor 集群彼此之间都可以证明对方拥有自己安全 key 的副本，而无须透露安全 key 去进行验证。这种身份认证机制使 Ceph 客户端可以和集群中的每个 OSD 直接通信，提高了效率，同时，安全 key 的扩展性也避免了其出现单点故障。

下面我们大概了解一下如何使用 cephx 认证系统。首先，客户端管理员必须设置用户，用户调用 ceph auth get-or-create-key 命令生成一个用户名和安全 key，cephx 认证系统将生成的用户名和安全 key 在 Monitor 集群中存储一个副本，并发送该用户的安全 key 给客户端管理员，这也就证明了客户端和 Monitor 集群共享一个安全 key，如下图所示。

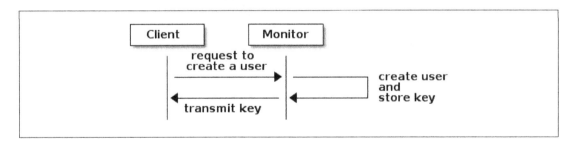

其次，在身份验证过程中，客户端传送用户名给 Monitor 集群，Monitor 集群会生成一个 session key，并将 session key 与该用户名关联的安全 key 一起加密，然后将加密信息传送给客户端。客户端使用共享的安全 key 对其解密并检索该 session key，该 session key 即可验证当前 session 的用户身份。客户端请求一个 ticket，Monitor 生成一个 ticket，并将其与用户的安全 key 一起加密，发送给客户端。客户端解密该 ticket 去对 OSD 进程和元数据服务器发起请求，如下图所示。

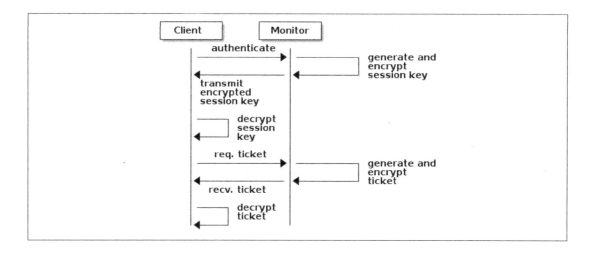

## 4.4.3 集群管理

Ceph 存储集群最重要的特点是自动管理、故障自动修复和智能 OSD 进程。下面介绍一下 Ceph 如何利用 CRUSH 算法实现数据存放、集群数据重平衡和自我故障恢复。

### 1．池化的概念

Ceph 存储集群支持池化，其可以理解为对象数据存储的一个逻辑概念。Ceph 客户端从 Monitor 进程中检索集群映射表，并将对象数据写入池中，如下图所示。池的大小或副本数量、CRUSH 规则设置和 PG 数量决定了 Ceph 如何存储对象数据。

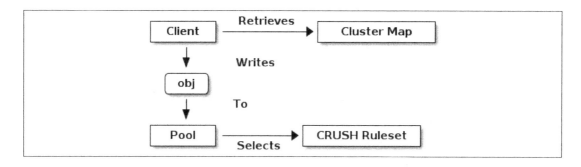

### 2．PG和OSD

每个池中包含多个 PG（Placement Group），CRUSH 可以动态地将 PG 映射给 OSD。当客户端需要存储数据时，CRUSH 会将对象数据映射给 PG，然后将 PG 映射给 OSD，完成对象数据的存储。

PG 位于对象数据和 OSD 之间，是一个中间层。如果 Ceph 客户端知道对象数据存放在哪个 OSD 上，则会在客户端和 OSD 之间建立一个联系，下次需要访问该对象数据时，可以直接访问。如果不知道，则 CRUSH 会将对象数据映射在 PG 上，再将该 PG 映射在 OSD 上，最终写在磁盘上。当 Ceph 存储集群增加或减少 OSD 时，Ceph 利用中间层 PG 进行数据的动态平衡，以保证存储具备最佳性能，如下图所示。

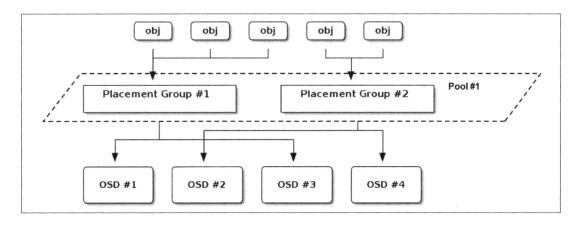

### 3．计算 PG ID

当 Ceph 客户端和 Monitor 建立联系后，客户端会对最新的集群映射表进行检索，由此可以得到关于 Monitor、OSD、元数据服务器的所有信息，但是仍不能获知关于对象数据存放位置的信息。对象数据的存放位置信息需要通过计算获得。

下面简要介绍一下计算对象数据存放位置的过程。当客户端需要存储对象数据时，CRUSH 会根据对象数据名称、Hash 码、PG 数量和池的名称计算出 PG ID，大概分为以下几个步骤。

（1）客户端输入池和对象数据的 ID 或者名称（例如，pool = "Liverpool"，object-id = "john"）。

（2）Ceph 将对象数据的 ID 进行 Hash 编码。

(3) Ceph 使用 Hash 编码和 PG 数量计算出 PG ID（例如，id=58）。

(4) Ceph 获知池 ID（例如，"Liverpool" = 4）。

(5) Ceph 将池 ID 和 PG ID 进行关联，即可确定对象数据的存放位置（例如，位置信息为 4.58）。

由上面的介绍可以得知，对于获知对象数据的存放位置信息，计算比在存储集群中检索速度更快。CRUSH 算法允许客户端计算对象数据的存放位置，并和 OSD 建立联系以存放和检索对象数据。

### 4．数据重平衡

当在 Ceph 存储集群中添加新的 OSD 时，CRUSH 会重新计算 PG ID，相应的集群映射表也会更新，基于重新计算的结果，对象数据的存放位置也会发生变化。下面简要介绍一下数据重平衡的过程：在一个含有两个 OSD 的存储集群中，在加入第三个 OSD 后，原有的两个 OSD 内的 PG 不会全部移动到新的 OSD 中，CRUSH 会重新计算，确定哪些 PG 需要迁移，大部分 PG 会保持原有配置和位置，少部分 PG 会进行迁移，如下图所示。重平衡的操作会使每个 OSD 释放出一些空闲空间，总体上增加了存储集群的容量。

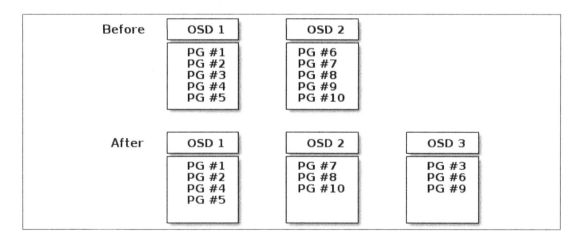

### 5．数据一致性

为了保证数据的一致性和清洁度，Ceph 存储集群会定期擦洗 PG 中的对象数据，大概原理就是 Ceph OSD 会将一个 PG 中的数据与另一个 PG 中的该数据副本进行比较，以此获知比对结果。数据擦洗分为两种：第一种是轻度擦洗，每天发生，可以捕捉到 OSD 的一些 Bug 或文件系统错误；第二种是重度擦洗，每周发生，在字节级别（bit-for-bit）进行数据比较，可以捕捉到一些硬盘坏块等错误，这种错误在轻度擦洗中无法检测到。

# 第 5 章 计量（Ceilometer）服务介绍

在 OpenStack 项目中，Ceilometer 计量服务组件以标准格式收集数据存储和系统资源的使用信息而进行统计和计费。同时，它也可以获取操作执行所产生的信息，触发通知。这些数据会被存储下来，作为计量数据。

随着计量服务组件收集数据量的不断增加，除了用于计费外，还可以有其他用途。例如，跟踪到某个系统负载压力过大而发出警告，可以据此对系统进行硬件资源扩容。

## 5.1　计量服务组件组成

Ceilometer 计量服务组件采用客户端部署 agent 的架构，收集数据，将数据存放在数据库中，或者提供一个处理传入请求的 API 服务。

Ceilometer 计量服务组件包括以下几部分。

- A compute agent（ceilometer-agent-compute）：运行在每个计算节点上，循环查询资源使用率统计情况。

- A central agent（ceilometer-agent-central）：运行在管理服务器上，循环查询资源使用率统计情况。

- A notification agent（ceilometer-agent-notification）：运行在管理服务器上，使用消息队列中的信息记录事件和计量数据。

- A collector（ceilometer-collector）：运行在管理服务器上，分发收集监测数据到数据存储上或展示给用户。

- An alarm evaluator（ceilometer-alarm-evaluator）：运行在管理服务器上，根据设置的阈值定义确定何时进行报警。

- An alarm notifier（ceilometer-alarm-notifier）：运行在管理服务器上，允许报警阈值在合适的范围内进行调整、设置。

- An API server（ceilometer-api）：运行在管理服务器上，提供数据访问。

以上这些服务通过 OpenStack 项目的消息总线进行通信。只有 Collector 和 API Server 可以访问数据存储。

## 5.2　计量服务组件支持列表

计量服务组件对 OpenStack 项目中的资源进行监控、统计和计费是有要求的，下面从数据库、虚拟化技术和网络三个方面分别介绍计量服务组件的支持情况。

### 1．数据库

计量服务组件最核心的部分就是数据库，它存储着事件、样例、定义警告触发的阈值和警告等。其所支持的数据库列表如下：

- ElasticSearch（仅支持事件存储）。

- MongoDB。

- MySQL。

- PostgreSQL。

- HBase。

## 2. 虚拟化技术

计算服务组件收集关于虚拟机的信息。其所支持的虚拟化技术列表如下：

- 以下虚拟化技术通过 Libvirt 支持：
  - KVM（Kernel-based Virtual Machine）。
  - QEMU（Quick Emulator）。
  - LXC（Linux Containers）。
  - UML（User-Mode Linux）。
- Hyber-V。
- Xen。
- VMware VSphere。

## 3. 网络

计量服务组件支持对 OpenStack 项目内部的 Neutron 网络服务组件和外部网络服务进行检索、计量。其支持列表如下：

- OpenStack 项目内部网络服务组件（Neutron）：
  - 基本网络模型
  - FWaaS（Firewall as a Service）。
  - LBaaS（Load Balance as a Service）。
  - VPNaaS（VPN as a Service）。
- 软件定义网络（SDN）：
  - OpenDayLight。
  - OpenContrail。

# 第 6 章 身份认证（Keystone）服务介绍

OpenStack 项目中的 Keystone 身份认证服务组件提供了认证、授权和目录的服务，其他 OpenStack 项目中的服务组件都需要使用它，彼此之间相互协作。当一个 OpenStack 服务组件收到用户的请求时，首先要交给 Keystone 身份认证服务组件检查该用户是否具有足够权限完成其提出的请求任务。Keystone 身份认证服务组件是唯一可以提供身份认证的服务组件。

Keystone 身份认证服务组件包括以下几部分。

- Server：使用一个程序接口提供认证和授权服务。
- Drivers：集成到服务器中，用作 OpenStack 项目之外和已经在项目之内（SQL Database）的程序访问身份认证信息。
- Modules：运行在使用认证服务的 OpenStack 服务组件中，监听服务请求，提取用户凭证，发送这些信息去服务器验证并对其授权。

当 Keystone 身份认证服务组件安装完毕后，需要将 OpenStack 项目中的每个服务组件都注册到其中，以使 Keystone 身份认证服务组件能够识别这些服务组件。

下面对 Keystone 身份认证服务组件中的一些概念进行介绍，以帮助在后续操作中可以更好地理解每个选项的含义。

### 1．Authentication

对用户进行身份确认的过程。对于一个操作请求，Keystone 身份认证服务组件会验证发起请求的用户所提供的凭证，该凭证既可以是用户名和密码，又可以是用户名和 API key。当 Keystone 身份认证服务组件验证用户凭证正确后，会发出一个认证令牌。在后续的请求中，用户会提供该认证令牌。

### 2．Credential

确认用户身份的数据，例如，用户名和密码、用户名和 API key、Keystone 身份认证服务组件提供的认证令牌。

### 3．Domain

项目和用户的集合，为认证实体定义了管理界限。域是个人、运营商或公司所拥有的空间，用户可以直接在域中进行管理操作。用户可以获得域管理员角色，创建项目、用户和组，分配角色给用户和组。

### 4．Endpoint

一个网络访问地址，通过它可以访问某个服务组件，以 URL 形式为主。

### 5．Group

组是域所拥有的所有用户的集合。组角色授予域或项目，应用于该组中所有的用户。从组中添加或删除用户会相应地获得或撤销该用户在域或项目中的角色和认证信息。

### 6．OpenStackClient

OpenStack 项目中服务组件的命令行接口。例如，用户可以使用 openstack service create 和 nova boot 命令注册服务或创建实例。

### 7．Project

集合或分离资源或对象的容器。根据不同的操作管理员，一个项目可以映射到客户、用户、组织或租户。

### 8．Region

OpenStack 部署的分区区域。可以关联多个子区域到一个区域，形成树形结构的大区域。虽然区域并不代表实际的物理位置，但是可以使用代表物理区域的名词进行命名，例如，ch-bj。

### 9．Role

定义一组用户的权利和权限，具备执行一系列指定操作的特性。Keystone 身份认证服务组件发送给用户的令牌环包含多个角色。当用户调用某个服务时，该服务会分析该用户的角色设置，确定该角色是否拥有操作和访问资源的权限。

### 10．Service

OpenStack 项目中的服务组件，例如，计算（Nova）、对象存储（Swift）等。这些服务组件提供一个 endpoint，用户通过该 endpoint 可以访问资源和执行操作。

### 11．Token

允许访问 OpenStack 项目 APIs 及各种资源的由字母和数字组成的字符串。认证令牌在任何时间都可以被撤销，同时也具备有效期限。在 OpenStack 的 L 版本中，Keystone 身份认证服务组件支持基于 Token 的验证，未来会支持更多的功能。Keystone 是一个集成的服务组件，具备很多功能和特性，而不是一个只具备身份认证和管理功能的技术方案。

### 12．User

使用 OpenStack 云服务的个人、系统或服务的一个数码表示凭证。Keystone 身份认证服务组件会确认用户发出请求的有效性。用户利用自身的 Token 可以登录和访问各种资源。

# 第 7 章 镜像（Glance）服务介绍

Glance 镜像服务组件是 OpenStack 项目的核心组件，它接收来自用户和 Nova 计算服务组件的 API 请求，支持在不同的介质上进行存储，包括 OpenStack 项目中的 Swift 对象存储。

Glance 镜像服务组件上运行的进程支持缓存、副本服务保证可用性和一致性，其他进程还包括审计、更新等服务。

 注意：

本书介绍的 Glance 镜像服务将虚拟机镜像放在控制节点的文件系统中，默认路径是 /var/lib/glance/images/。在进行下面的操作之前，必须确认该文件系统拥有足够的空闲空间用以存储虚拟机镜像。

Glance 镜像服务组件包括以下几部分。

- glance-api：接收发现、检索和存储镜像的 API 请求。
- glance-registry：存储、处理和检索镜像元数据。镜像元数据包括大小、类型等。

 警告：

glance-registry 是 Glance 镜像服务组件内部的私有服务，不能开放给外部用户使用。

- Database:存储镜像元数据。Glance 镜像服务组件支持多种数据库,大部分使用 MySQL 或者 SQLite 数据库。
- Storage repository for image files:虚拟机镜像文件的存储介质支持多种类型,包括普通的文件系统、对象存储、分布式存储(如 Ceph)、HTTP 和 Amazon S3。在使用这些存储介质时,务必确认存储介质是否只支持只读使用。

# 第 8 章
# 仪表板（Horizon）服务介绍

OpenStack 项目中的 Horizon 仪表板服务组件是一个模块化应用，提供基于 Web 的图形界面，允许用户管理各种计算资源和服务组件，如下图所示。

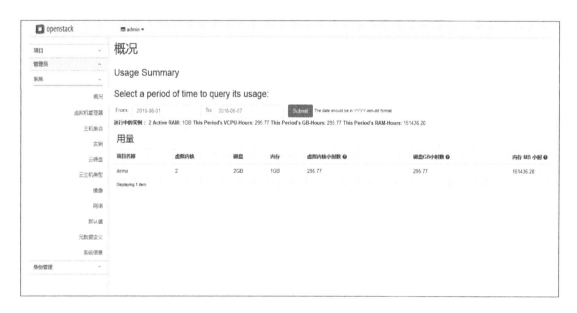

Horizon 仪表板服务组件通常使用 Apache 应用服务器的 mod_wsgi 进行部署，可以根据自己的实际需求对 Horizon 仪表板服务组件的代码进行修改。在第 2 篇中，我们会详细

## 第 8 章 仪表板（Horizon）服务介绍

介绍如何安装部署 Horizon 仪表板服务组件。在第 3 篇中，我们会概述如何修改关于 Horizon 仪表板服务组件的代码。

从网络架构角度看，用户和 OpenStack 项目中的所有服务组件都必须可以访问 Horizon 仪表板服务组件，这样 Horizon 仪表板才能提供服务给用户与发送请求给其他服务组件。

在 Horizon 仪表板服务组件中，集成了各服务组件的功能，如下。

- 项目-计算-概况，如下图所示。

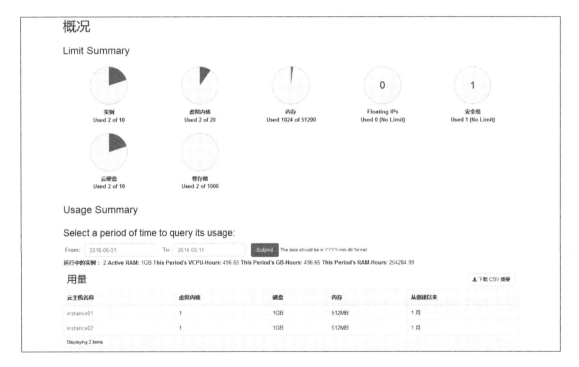

- 项目-计算-实例，如下图所示。

- 项目-计算-云硬盘，如下图所示。

- 项目-计算-镜像，如下图所示。

- 项目-计算-访问&安全，如下图所示。

- 项目-网络-网络拓扑，如下图所示。

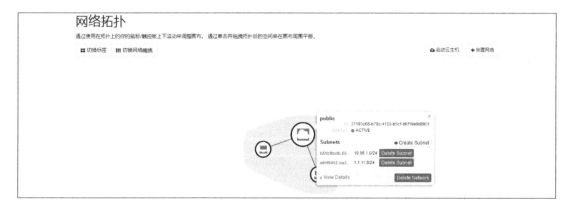

- 项目-网络-网络，如下图所示。

- 管理员-系统-概况，如下图所示。

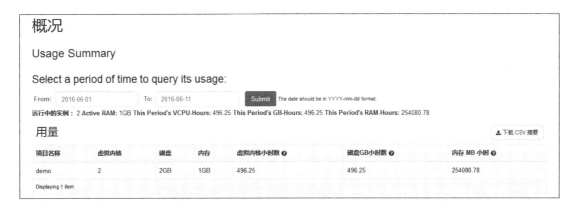

- 管理员-系统-所有虚拟机管理器，如下图所示。

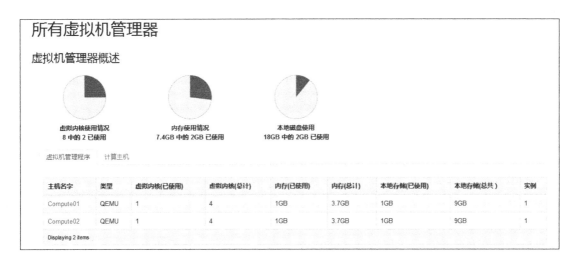

- 管理员-系统-主机集合，如下图所示。

- 管理员-系统-实例，如下图所示。

- 管理员-系统-云硬盘，如下图所示。

- 管理员-系统-云主机类型，如下图所示。

- 管理员-系统-镜像，如下图所示。

- 管理员-系统-网络，如下图所示。

- 管理员-系统-系统信息，如下图所示。

- 身份管理-域，如下图所示。

- 身份管理-项目，如下图所示。

- 身份管理-用户，如下图所示。

- 身份管理-角色，如下图所示。

# 第 9 章
# 编排（Heat）服务介绍

Heat 编排服务组件提供基于模板的服务，它集成 OpenStack 项目中的许多核心服务组件于一个模板中，使用这个模板可以创建许多类型的资源，如虚拟机实例、Floating IPs、卷设备、安全组和用户等。当然，还有一些其他高级功能，如实现虚拟机实例高可用、虚拟机实例的自动横向扩展等。这些功能使 OpenStack 项目适用的应用场景更多，获得了更多的用户群。

Heat 编排服务组件包括以下几部分。

- heat command-line client：提供命令行客户端，与 heat-api 交互通信以运行 AWS CloudFormation APIs。开发人员可以直接使用 Heat 编排服务组件 REST API。
- heat-api component：通过使用 RPC 将请求发送到 heat-engine 处理 API 请求。
- heat-api-cfn component：与 AWS CloudFormation 兼容的 AWS Query API，通过使用 RPC 将请求发送到 heat-engine 处理这些 API 请求。
- heat-engine：编排模板的运行，并提供日志、事件输出给管理员。

随着 Heat 编排服务组件的不断发展，它已经开始兼容 AWS CloudFormation 模板，许多 AWS CloudFormation 模板可以直接在 OpenStack 环境中运行。

# 第 2 篇

# 安装配置篇

- 第 10 章 OpenStack 安装配置准备
- 第 11 章 身份认证（Keystone）服务安装配置
- 第 12 章 镜像（Glance）服务安装配置
- 第 13 章 计算（Nova）服务安装配置
- 第 14 章 网络（Neutron）服务安装配置
- 第 15 章 仪表板（Horizon）服务安装配置
- 第 16 章 块存储（Cinder）服务安装配置
- 第 17 章 对象存储（Swift）服务安装配置
- 第 18 章 编排（Heat）服务安装配置
- 第 19 章 计量（Ceilometer）服务安装配置
- 第 20 章 建立虚拟机实例测试

# 第 10 章
# OpenStack 安装配置准备

本书将以 OpenStack Liberty 版本为例进行安装部署，物理主机节点的操作系统选择 SUSE 系统（openSUSE 13.2 或 SUSE Linux Enterprise Server 12）。OpenStack 由多个服务组件构成，既可以选择全部服务组件，也可以选择部分服务组件进行安装部署。

在进行 OpenStack 安装和配置之前，需要仔细思考以下几个问题：

- 我们需要使用哪几个服务组件，使我们的架构达到最优，既不臃肿又不单薄？
- 我们需要使用哪些方法和策略保证架构安全？
- 我们需要使用哪些工具或者技术使 OpenStack 的安装部署自动化，且管理简单？

## 10.1 架构设计

本书中的 OpenStack 示例环境架构设计包括 5 个主机节点，分别为一个控制节点、两个计算和块存储节点（计算和存储服务在一个节点上）和两个对象存储节点。

### 1. 控制节点

控制节点非常重要，许多服务组件和软件运行其上，如 Horizon 仪表板服务组件、

Keystone 身份认证服务组件、Glance 镜像服务组件，以及支持 OpenStack 运行的各种软件，包括数据库、消息队列和时钟服务等。

控制节点要求最少有两个网络端口，以满足需要：一个是管理/数据网络，用于主机节点间的数据传输和网络连接；另一个是公共网络，用于虚拟机实例的网络连接。

### 2．计算节点

计算节点运行着虚拟化软件（默认使用 KVM 虚拟化技术）和网络服务，用于提供虚拟机实例及其网络连接和防火墙服务（通过安全组提供）。

计算节点要求最少有两个网络端口，作用与控制节点相同。在 OpenStack 环境中可以部署多个计算节点，以增加整个 OpenStack 环境的计算资源的容量和体量。

### 3．块存储节点

块存储服务不是必选服务，为虚拟机实例提供块存储服务。在真正的生产环境中，管理网络（存储节点与控制节点通信）和数据传输网络（存储节点和计算节点通信）必须是相互独立的，以增强整个架构的性能和安全。

块存储节点要求最少有一个网络端口，用于提供数据传输。但是可以部署多个存储节点，以增加整个环境的存储容量。

### 4．对象存储节点

对象存储服务不是必选服务，为各种服务组件提供存储服务。在真正的生产环境中，管理网络（存储节点与控制节点通信）和数据传输网络（存储节点和计算节点通信）必须是独立的，以增强整个架构的性能和安全。

对象存储至少包括两个存储节点，每个节点要求最少有一个网络端口。

### 5．网络节点

网络节点可以部署多种网络拓扑模型，具体如下。

1）Flat 模式

Flat 模式是一种简单的网络服务形式，在 OSI 模型中属于第二层（数据链路层），虚拟网络和物理网络进行桥接，路由交换依赖于物理网络中的 OSI 模型第三层（网络层），DHCP 服务提供了对虚拟机实例的 IP 地址分配功能。

该模式缺少对私有网络的支持，不能在 OpenStack 环境内部实现网络自服务功能、基于 OSI 模型三层的虚拟路由功能和其他服务，如 LBaaS 和 FWaaS 等。

网络模式 1：Flat 模式，如下图所示。

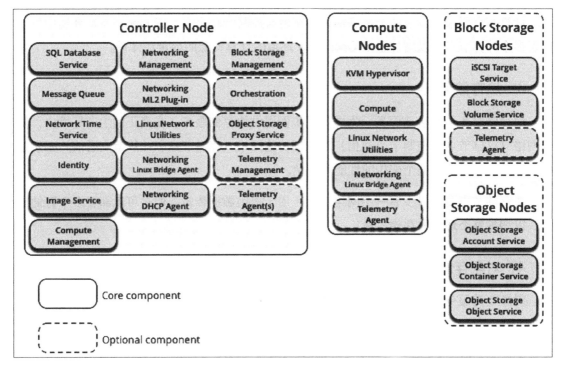

2）VLAN、VXLAN 模式

该模式可以提供网络自服务和路由交换，使用 NAT 技术实现数据在虚拟网络和物理网络之间的路由连接，同时提供了各种其他服务，如 LBaaS 和 FWaaS 等。

网络模式 2：VLAN 和 VXLAN 模式，如下图所示。

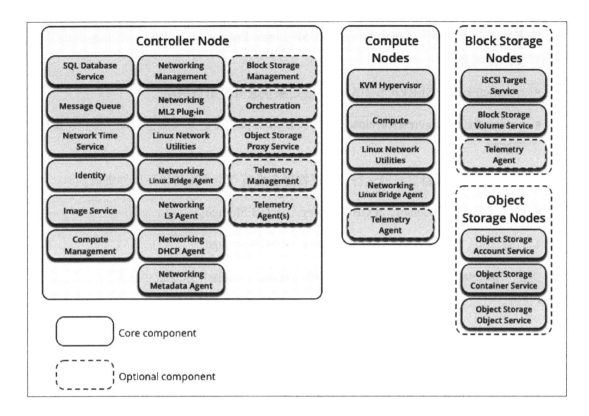

## 10.2 基础环境准备

本章主要介绍在安装配置 OpenStack 各服务组件之前，如何准备基础环境。基础环境准备是否充分对后续的安装配置是否成功至关重要。整个 OpenStack 环境需要仪表板服务（Horizon）、计算服务（Nova）、身份认证服务（Keystone）、网络服务（Neutron）、镜像服务（Glance）、块存储（Cinder）、对象存储（Swift）、计量服务（Ceilometer）、编排服务（Heat）。

> **注意：**
> 在后续的介绍中，将默认使用各服务组件的英文名称，这样既简单又有利于后续学习。随着学习的深入，就会慢慢发现，有效的信息大多是以英文出现的。

需要配置的所有服务组件都要运行在 Linux 操作系统上，同时需要拥有管理员权限的用户，既可以是 root 用户，也可以是有 sudo 权限的普通用户。

本书实例使用的操作系统为 SLES（SUSE Linux Enterprise Server），使用 systemctl 命令可以操作各项服务。

- systemctl enable SERVICE_NAME：设定某项服务开机自启动。
- systemctl start/stop/restart/reload SERVICE_NAME：设定某项服务启动、停止、重启、重加载。

为使 OpenStack 项目更好地运行，建议物理主机节点至少配置以下硬件资源。

- 控制节点：1 Processor，4 GB Memory，20 GB Storage。
- 计算节点：2 Processor，5 GB Memory，40 GB Storage。

为使 OpenStack 项目能够使用更多的硬件资源，建议操作系统最小化安装，并且必须是 64 位的。

每个物理主机节点必须有一个独立的磁盘分区安装操作系统，同时在计算节点上必须有一块独立磁盘分区作块存储使用。

基于学习的需要，可以采取虚拟机作为物理主机节点建立 OpenStack 云环境，其优势和要求如下：

- 一个物理机可以支持多个虚拟机，用于安装多个服务组件，成本最小化。
- 目前市场上主流的虚拟化产品（VMware WorkStation、VirtualBox 等）都支持 SnapShot 功能，方便在配置出现问题时进行回退。
- 物理磁盘数量和大小及网卡数量可以随意调整和定义。

当然，使用虚拟机作为物理主机节点安装 OpenStack 的服务组件也有缺点，如性能降低、缺少硬件加速等功能。

 注意：

  如果确定使用虚拟机作为物理主机节点部署 OpenStack，请确认物理 CPU 是否支持虚拟化功能、网卡是否支持 MAC 地址过滤功能。

## 10.2.1 安全设置规则

OpenStack 项目的各服务组件支持不同类型的安全规则，如密码、策略和加密等，有的还支持数据库加密和消息代理，但是有一个共性，那就是都支持设置密码作为安全规则。

为了简化整个 OpenStack 环境的安装部署过程，我们对所有的服务组件都使用密码作为安全规则。

下表是 OpenStack 项目各服务组件设置的密码格式和规则。

| 密码格式 | |
| --- | --- |
| 密 码 | 解 释 |
| Database password | 数据库密码 |
| ADMIN_PASS | admin 用户密码 |
| CEILOMETER_DBPASS | Ceilometer 服务组件的数据库密码 |
| CEILOMETER_PASS | Ceilometer 服务组件的用户密码 |
| CINDER_DBPASS | Cinder 服务组件的数据库密码 |
| CINDER_PASS | Cinder 服务组件的用户密码 |
| DASH_DBPASS | Horizon 仪表板服务组件的数据库密码 |
| DEMO_PASS | demo 用户密码 |
| GLANCE_DBPASS | Glance 服务组件的数据库密码 |
| GLANCE_PASS | Glance 服务组件的用户密码 |
| HEAT_DBPASS | Heat 服务组件的数据库密码 |
| HEAT_DOMAIN_PASS | Heat 服务组件的域密码 |
| HEAT_PASS | Heat 服务组件的用户密码 |
| KEYSTONE_DBPASS | Keystone 服务组件的数据库密码 |
| NEUTRON_DBPASS | Neutron 服务组件的数据库密码 |
| NEUTRON_PASS | Neutron 服务组件的用户密码 |
| NOVA_DBPASS | Nova 服务组件的数据库密码 |
| NOVA_PASS | Nova 服务组件的用户密码 |

续表

| 密码格式 | |
|---|---|
| 密　　码 | 解　　释 |
| RABBIT_PASS | RabbitMQ 的密码 |
| SWIFT_PASS | Swift 服务组件的用户密码 |

## 10.2.2　主机节点网络设置

在主机节点上安装完操作系统后，需要为每台主机节点配置网络。建议关闭 OpenStack 环境中有关网络的所有自动管理工具，手动编辑有关网络连接的配置文件。

由于 OpenStack 环境中所有的主机节点都需要安装或更新软件包，进行 DNS 解析和 NTP 同步，所以 OpenStack 环境中的所有主机节点必须能够连接互联网。

网络布局如下图所示。

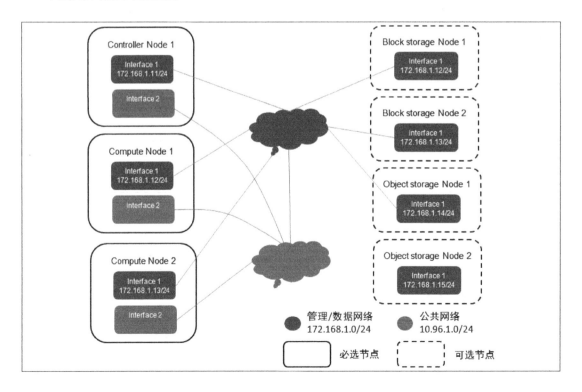

该网络布局有以下几点需要注意和解释：

- 可以根据实际情况修改以上网络信息，包括 IP 地址和网关。
- 管理/数据网络可以提供主机节点间的数据传输，也可以提供互联网访问。
- 公共网络为虚拟机实例提供网络连接。
- 每个主机节点的 IP 地址与主机名必须能够相互解析。

**注意**：
在安装操作系统时，建议关闭防火墙，以防在后续配置过程中产生各种问题。

### 1. 控制节点配置

1）网络配置

首先配置管理/数据网络的网卡信息。

IP 地址：172.168.1.11。

子网掩码：255.255.255.0。

默认网关：172.168.1.254。

公共网络的网卡不需要配置 IP 地址，编辑 /etc/sysconfig/network/ifcfg-INTERFACE_NAME 文件，添加以下信息：

```
STARTMODE='auto'
BOOTPROTO='static'
```

按照以上修改完成后，可以重启机器或者重启网络服务生效。

```
# reboot
或
# systemctl restart network.service
```

2）名字解析

首先为控制节点设置主机名，然后编辑 /etc/hosts 文件。配置如下：

```
#controller
```

```
172.168.1.11     Controller

#compute01
172.168.1.12     Compute01

#compute02
172.168.1.13     Compute02

#NTPServer
172.168.1.1      NTPServer

#Object-sto01
172.168.1.14     Object_sto01

#Object-sto02
172.168.1.15     Object_sto02
```

> **警告**：
> Linux 操作系统版本的不同可能导致/etc/hosts 文件内容的不同，一定要注释或者删除影响名字解析的内容（如 127.0.1.1）。但是不要删除 127.0.0.1 的名字解析。

> **注意**：
> 为了减少部署的复杂性和增加部署的完整性，该文件（/etc/hosts）包含了所有主机节点的名字解析内容，即使可能不会使用的节点信息（如 Object 存储节点部分）也包含其中。

## 2. 计算节点配置

1）网络配置

首先配置管理/数据网络的网卡信息。

IP 地址：172.168.1.12/13（第一个节点 IP 是 172.168.1.12，第二个节点 IP 是 172.168.1.13）

子网掩码：255.255.255.0。

默认网关：172.168.1.254。

公共网络的网卡不需要配置 IP 地址,编辑/etc/sysconfig/network/ifcfg-INTERFACE_NAME 文件,添加以下信息:

```
STARTMODE='auto'
BOOTPROTO='static'
```

按照以上修改完成后,可以重启机器或者重启网络服务生效。

```
#reboot
或
# systemctl restart network.service
```

2) 名字解析

首先为计算节点设置主机名,然后编辑 /etc/hosts 文件。配置如下:

```
#controller
172.168.1.11    Controller

#compute01
172.168.1.12    Compute01

#compute02
172.168.1.13    Compute02

#NTPServer
172.168.1.1     NTPServer

#Object-sto01
172.168.1.14    Object_sto01

#Object-sto02
172.168.1.15    Object_sto02
```

## 3. 块存储节点配置(可选)

由于块存储节点和计算节点在同一节点上,关于块存储节点的配置已经在计算节点配置部分完成,在此不再赘述。

## 4．对象存储节点配置（可选）

1）网络配置

首先配置管理/数据网络的网卡信息。

IP 地址：172.168.1.14/15（第一个节点 IP 是 172.168.1.14，第二个节点 IP 是 172.168.1.15）。

子网掩码：255.255.255.0。

默认网关：172.168.1.254。

按照以上修改完后，可以重启机器或者重启网络服务生效。

```
#reboot
或
# systemctl restart network.service
```

2）名字解析

首先为对象存储节点设置主机名，然后编辑 /etc/hosts 文件。配置如下：

```
#controller
172.168.1.11    Controller

#compute01
172.168.1.12    Compute01

#compute02
172.168.1.13    Compute02

#NTPServer
172.168.1.1     NTPServer

#Object-sto01
172.168.1.14    Object_sto01

#Object-sto02
172.168.1.15    Object_sto02
```

## 5. 检查配置

以上各主机节点信息配置完成后，务必检查并确认所有主机节点均可以访问互联网和相互访问。

- 在控制节点测试是否可以访问互联网。

```
# ping -c 4 openstack.org
PING openstack.org (162.242.140.107) 56(84) bytes of data.
64 bytes from 162.242.140.107: icmp_seq=1 ttl=37 time=204 ms
64 bytes from 162.242.140.107: icmp_seq=2 ttl=37 time=204 ms
64 bytes from 162.242.140.107: icmp_seq=3 ttl=37 time=204 ms
64 bytes from 162.242.140.107: icmp_seq=4 ttl=37 time=204 ms

--- openstack.org ping statistics ---
4 packets transmitted, 4 received, 0% packet loss, time 3001ms
rtt min/avg/max/mdev = 204.213/204.483/204.634/0.479 ms
```

- 在控制节点测试是否可以访问计算节点。

```
# ping -c 4 Compute01
PING Compute01 (172.168.1.12) 56(84) bytes of data.
64 bytes from Compute01 (172.168.1.12): icmp_seq=1 ttl=64 time=0.606 ms
64 bytes from Compute01 (172.168.1.12): icmp_seq=2 ttl=64 time=0.530 ms
64 bytes from Compute01 (172.168.1.12): icmp_seq=3 ttl=64 time=0.552 ms
64 bytes from Compute01 (172.168.1.12): icmp_seq=4 ttl=64 time=0.566 ms

--- Compute01 ping statistics ---
4 packets transmitted, 4 received, 0% packet loss, time 3000ms
rtt min/avg/max/mdev = 0.530/0.563/0.606/0.036 ms
```

- 在计算节点测试是否可以访问互联网。

```
# ping -c 4 openstack.org
PING openstack.org (162.242.140.107) 56(84) bytes of data.
64 bytes from 162.242.140.107: icmp_seq=1 ttl=37 time=206 ms
64 bytes from 162.242.140.107: icmp_seq=2 ttl=37 time=205 ms
64 bytes from 162.242.140.107: icmp_seq=3 ttl=37 time=205 ms
64 bytes from 162.242.140.107: icmp_seq=4 ttl=37 time=205 ms
```

```
--- openstack.org ping statistics ---
4 packets transmitted, 4 received, 0% packet loss, time 3003ms
rtt min/avg/max/mdev = 205.258/205.682/206.132/0.551 ms
```

- 在计算测试是否可以访问控制节点。

```
# ping -c 4 Controller
PING Controller (172.168.1.11) 56(84) bytes of data.
64 bytes from Controller (172.168.1.11): icmp_seq=1 ttl=64 time=0.344 ms
64 bytes from Controller (172.168.1.11): icmp_seq=2 ttl=64 time=0.613 ms
64 bytes from Controller (172.168.1.11): icmp_seq=3 ttl=64 time=0.576 ms
64 bytes from Controller (172.168.1.11): icmp_seq=4 ttl=64 time=1.03 ms

--- Controller ping statistics ---
4 packets transmitted, 4 received, 0% packet loss, time 3003ms
rtt min/avg/max/mdev = 0.344/0.641/1.034/0.250 ms
```

> 注意：
> 所有 Linux 操作系统都有严格的防火墙规则，在安装部署过程中，可能需要修改防火墙规则。基于测试和学习的目的，为了避免操作错误和繁杂，建议关闭防火墙。

## 10.2.3 节点时钟同步

整个 OpenStack 环境中所有主机节点的时间必须是相同的，因此需要安装一个时间同步软件 Chrony，用于在节点间实现时钟同步。建议选择控制节点作为时钟同步服务器（如果使用虚拟机作为物理主机节点部署 OpenStack 环境，则选择宿主机作为时钟同步服务器），既可以获得良好的网络时延性，又能保证思维逻辑的正确性。

### 1. 控制节点安装和配置

（1）使用 zypper 命令安装 Chrony。

```
# zypper addrepo -f obs://network:time/SLE_12 network_time
# zypper refresh
# zypper install chrony
```

(2) 编辑/etc/chrony.conf 文件，添加、修改和删除内容以匹配当前的环境。

```
server    NTPServer    iburst
```

使用合适的主机名或者 IP 地址替换 NTPServer，文件中可以配置多个时钟同步服务器。

(3) 启动 Chrony 服务，并配置 Chrony 服务开机自启动。

```
# systemctl enable chronyd.service
# systemctl start chronyd.service
```

### 2．其他节点安装和配置

(1) 使用 zypper 命令安装 Chrony。

```
# zypper addrepo -f obs://network:time/SLE_12 network_time
# zypper refresh
# zypper install chrony
```

(2) 编辑 /etc/chrony.conf 文件，添加、修改和删除内容以匹配当前的环境。

```
server    NTPServer    iburst
```

使用合适的主机名或者 IP 地址替换 NTPServer，文件中可以配置多个时钟同步服务器。

(3) 启动 Chrony 服务，并配置 Chrony 服务开机自启动。

```
# systemctl enable chronyd.service
# systemctl start chronyd.service
```

### 3．检查配置

以上全部配置完成后，务必检查并确认所有的主机节点都已经完成时钟同步。

(1) 检查并确认控制节点是否完成时钟同步。

```
# chronyc sources
210 Number of sources = 1
MS Name/IP address         Stratum Poll Reach LastRx Last sample
===============================================================================
^? NTPServer                  1   10   377    154  -120.7s[-120.7s] +/-  10.9s
```

(2)检查并确认其他节点是否完成时钟同步。

```
# chronyc sources
210 Number of sources = 1
MS Name/IP address         Stratum Poll Reach LastRx Last sample
===============================================================================
^? NTPServer                  1   10    377    601   -120.5s[-120.5s] +/-  10.9s
```

### 10.2.4 配置 OpenStack 安装源和运行环境

在运行以下操作之前，一定要确认 OpenStack 环境中主机节点的 Linux 操作系统已经正确安装完毕。以下步骤需要在所有主机节点上运行。

 **注意**：

由于操作系统中的自动更新会影响 OpenStack 的正常运行，所以务必关闭操作系统的自动更新功能。

（1）配置 OpenStack 安装源。

```
# zypper addrepo -f obs://Cloud:OpenStack:Liberty/SLE_12 Liberty
```

（2）更新操作系统软件包。如果包含操作系统 Kernel 升级，则必须重启操作系统使其生效。

```
# zypper refresh && zypper dist-upgrade
```

（3）安装 OpenStack 运行环境软件包。

```
# zypper install python-openstackclient
```

### 10.2.5 安装和配置 SQL 数据库

OpenStack 中的大部分服务组件都需要使用 SQL 数据库，它们通常运行在控制节点上。本书示例我们使用 MariaDB，当然也可以使用 MySQL 或者 PostgreSQL，根据操作系统不同可以选择不同的数据库。

(1) 安装 MariaDB。

```
# zypper install mariadb-client mariadb python-PyMySQL
```

(2) 创建并编辑/etc/my.cnf.d/mariadb_openstack.cnf 文件。在文件的[mysqld]部分，设置 bind-address 地址为控制节点的 IP 地址，使其他节点能够通过管理/数据网络与其通信。设置正确的字符编码规则。

```
[mysqld]
bind-address = 172.168.1.11
default-storage-engine = innodb
innodb_file_per_table
collation-server = utf8_general_ci
character-set-server = utf8
```

(3) 启动数据库服务并设置为开机自启动。

```
# systemctl enable mysql.service
# systemctl start mysql.service
```

## 10.2.6 安装和配置 NoSQL 数据库

由于 Ceilometer 计量服务组件需要使用 NoSQL 数据库，所以我们选择在控制节点安装 MongoDB。

 注意：

NoSQL 数据库是 Ceilometer 计量服务组件必需的。如果不使用 Ceilometer 服务，则可以不安装 NoSQL 数据库。

(1) 在操作系统上配置 MongoDB 安装源。

```
# zypper addrepo -f obs://server:database/SLE_12 Database
```

(2) 安装 MongoDB。

```
# zypper install mongodb
```

(3) 编辑 /etc/mongodb.conf 文件，增加或者修改以下内容。

```
bind_ip = 172.168.1.11,::1
smallfiles = true
```

MongoDB 数据库默认在 /var/lib/mongodb/journal 目录中创建一些 1GB 大小的日志文件。为了避免单个日志文件过大和使用过多的文件系统空间，可以打开 smallfiles 参数，使每个日志文件大小控制在 128MB，并且文件系统空间最多使用 512MB。

（4）启动 MongoDB 服务并设置为开机自启动。

```
# systemctl enable mongodb.service
# systemctl start mongodb.service
```

## 10.2.7　安装和配置消息队列

OpenStack 中的多个服务组件间需要互相协调运作和更新状态信息，因此需要有一个消息队列完成这些工作。OpenStack 支持多种消息队列软件，如 RabbitMQ、Qpid 和 ZeroMQ，消息队列一般运行在控制节点上。由于大部分 OpenStack 版本都支持 RabbitMQ，所以本次试验我们使用 RabbitMQ。如果使用其他类型的消息队列，则务必确认 OpenStack 版本与消息队列软件的兼容性。

（1）安装 RabbitMQ。

```
# zypper install rabbitmq-server
```

（2）启动 RabbitMQ 并设置为开机自启动。

```
# systemctl enable rabbitmq-server.service
# systemctl start rabbitmq-server.service
```

如果 RabbitMQ 不能正常启动，并且报错"nodedown"，请执行以下操作：

- 复制文件/usr/lib/systemd/system/epmd.socket 到/etc/systemd/system 目录。

- 编辑文件/etc/systemd/system/epmd.socket，增加以下内容。

```
[Socket]
ListenStream=MANAGEMENT_INTERFACE_IP_ADDRESS:4369
```

替换 MANAGEMENT_INTERFACE_IP_ADDRESS 为控制节点的管理/数据 IP 地址。

- 重新启动 RabbitMQ。

（3）在 RabbitMQ 中创建用户。

```
# rabbitmqctl add_user openstack RABBIT_PASS
```

替换 RABBIT_PASS 为合适的密码。

 **注意：**
　　一定要记录好将要或已经创建的用户及其密码。在后续操作中，会频繁使用这些信息。

（4）设置 OpenStack 用户可以配置、写和读 RabbitMQ。

```
# rabbitmqctl set_permissions openstack ".*" ".*" ".*"
```

# 第 11 章 身份认证（Keystone）服务安装配置

Keystone 身份认证服务组件是 OpenStack 项目中默认使用的身份认证管理系统。安装完 Keystone 身份认证服务组件后，可以通过/etc/keystone/keystone.conf 文件进行配置。在设计架构允许的情况下，可以配置独立的日志记录文件，方便问题跟踪与解决。

## 11.1 安装和配置

本节详细描述如何在控制节点上安装和配置 Keystone 身份认证服务组件。

（1）创建数据库。

① 在操作系统终端连接数据库。

```
# mysql -u root -p
```

因为我们使用 root 用户登录，所以在提示输入密码时，按回车键即可建立连接。

② 创建 Keystone 数据库。

```
MariaDB [(none)]> CREATE DATABASE keystone;
```

③ Keystone 数据库的访问权限设置。

```
GRANT ALL PRIVILEGES ON keystone.* TO 'keystone'@'localhost' \
  IDENTIFIED BY 'KEYSTONE_DBPASS';
GRANT ALL PRIVILEGES ON keystone.* TO 'keystone'@'%' \
  IDENTIFIED BY 'KEYSTONE_DBPASS';
```

替换 KEYSTONE_DBPASS 为合适的密码。

④ 退出数据库。

```
MariaDB [(none)]> exit
```

（2）生成一串随机值作为管理 Token。

```
# openssl rand -hex 10
```

（3）安装 Apache HTTP Server 和 MemCached。

```
# zypper install openstack-keystone apache2-mod_wsgi \
  memcached python-python-memcached
```

（4）启动 MemCached 并设置为开机自启动。

```
# systemctl enable memcached.service
# systemctl start memcached.service
```

（5）修改文件 /etc/keystone/keystone.conf 并增加以下内容。

① 在 [DEFAULT] 项，定义管理令牌值。

```
[DEFAULT]
...
admin_token = ADMIN_TOKEN
```

使用第（2）步生成的管理令牌值替换 ADMIN_TOKEN。

② 在 [database] 项，配置数据库连接。

```
[database]
...
connection = mysql+pymysql://keystone:KEYSTONE_DBPASS@Controller/ keystone
```

使用已经定义的 Keystone 数据库密码替换 KEYSTONE_DBPASS。

③ 在 [memcache] 项，配置 MemCached 服务。

```
[memcache]
...
servers = localhost:11211
```

④ 在[token]项，增加以下内容。

```
[token]
...
provider = uuid
driver = memcache
```

⑤ 在[revoke] 项，增加以下内容。

```
[revoke]
...
driver = sql
```

⑥（可选）在[DEFAULT] 项，打开日志功能，方便问题处理。

```
[DEFAULT]
...
verbose = True
```

（6）将 Keystone 身份认证服务信息同步到 Keystone 的数据库中。

```
# su -s /bin/sh -c "keystone-manage db_sync" keystone
```

 错误：

执行此步骤时，报错"error : CRITICAL keystone [-] OperationalError: (pymysql.err.OperationalError) (1045, u"Access denied for user 'keystone'@'Controller' (using password: YES)" )"。

从错误信息分析，大概是权限问题。仔细分析，发现进行信息同步是 Keystone 用户与 Controller 服务器进行交互，在进行数据库访问权限设置时，并没有 Controller，因而问题可能存在于此。对 Controller 设置以下权限：

```
mysql -u -root -p
  GRANT ALL PRIVILEGES ON keystone.* TO 'keystone'@' Controller' \
IDENTIFIED BY 'KEYSTONE_DBPASS';
```

再次执行第（6）步，成功。

（7）编辑/etc/sysconfig/apache2 文件，设置 APACHE_SERVERNAME 为控制节点名字（与/etc/hosts 定义的控制节点名字一致）。

```
APACHE_SERVERNAME="Controller"
```

（8）创建文件/etc/apache2/conf.d/wsgi-keystone.conf，并添加以下内容。

```
Listen 5000
Listen 35357

<VirtualHost *:5000>
    WSGIDaemonProcess keystone-public processes=5 threads=1 user=keystone group=keystone display-name=%{GROUP}
    WSGIProcessGroup keystone-public
    WSGIScriptAlias / /usr/bin/keystone-wsgi-public
    WSGIApplicationGroup %{GLOBAL}
    WSGIPassAuthorization On
    <IfVersion >= 2.4>
      ErrorLogFormat "%{cu}t %M"
    </IfVersion>
    ErrorLog /var/log/apache2/keystone.log
    CustomLog /var/log/apache2/keystone_access.log combined

    <Directory /usr/bin>
       <IfVersion >= 2.4>
          Require all granted
       </IfVersion>
       <IfVersion < 2.4>
          Order allow,deny
          Allow from all
       </IfVersion>
    </Directory>
</VirtualHost>

<VirtualHost *:35357>
    WSGIDaemonProcess keystone-admin processes=5 threads=1 user=keystone group=keystone display-name=%{GROUP}
    WSGIProcessGroup keystone-admin
```

```
    WSGIScriptAlias / /usr/bin/keystone-wsgi-admin
    WSGIApplicationGroup %{GLOBAL}
    WSGIPassAuthorization On
    <IfVersion >= 2.4>
      ErrorLogFormat "%{cu}t %M"
    </IfVersion>
    ErrorLog /var/log/apache2/keystone.log
    CustomLog /var/log/apache2/keystone_access.log combined

    <Directory /usr/bin>
        <IfVersion >= 2.4>
            Require all granted
        </IfVersion>
        <IfVersion < 2.4>
            Order allow,deny
            Allow from all
        </IfVersion>
    </Directory>
</VirtualHost>
```

（9）修改 /etc/keystone 目录及该目录下所有文件的属主。

```
# chown -R keystone:keystone /etc/keystone
```

（10）激活 Apache 组件 mod_version。

```
# a2enmod version
```

（11）启动 Apache HTTP 服务并设置为开机自启动。

```
# systemctl enable apache2.service
# systemctl start apache2.service
```

## 11.2　创建 service entity 和 API endpoint

　　Keystone 身份认证服务组件为 OpenStack 项目中其他服务组件的使用和操作提供了服务目录，任何一个 OpenStack 项目中的服务组件都需要注册一个 service entity 和提供多个 API endpoint。

### 1．service entity

OpenStack 中有多个 service entity，如 Nova 计算服务组件、Cinder 块存储服务组件、Glance 镜像服务组件。它们提供多个 endpoint，用户可以通过 endpoint 访问资源和执行操作。

### 2．API endpoint

API endpoint 可以是某个进程或者服务，用户可以通过这些进程或者服务访问 OpenStack 项目中服务组件的 API。

## 11.2.1 准备

首先配置用户操作的环境变量。

（1）配置 admin_token。

```
# export OS_TOKEN=ADMIN_TOKEN
```

使用在 11.1 节生成的 ADMIN_TOKEN 值替换这里的 ADMIN_TOKEN。

（2）配置 endpoint 的通信地址。

```
# export OS_URL=http://controller:35357/v3
```

（3）配置 Keystone 身份认证服务组件的 API 版本。

```
# export OS_IDENTITY_API_VERSION=3
```

## 11.2.2 创建过程

### 1．创建Keystone身份认证服务组件的service entity

```
# openstack service create \
  --name keystone --description "OpenStack Identity" identity
+-------------+----------------------------------+
| Field       | Value                            |
+-------------+----------------------------------+
| description | OpenStack Identity               |
| enabled     | True                             |
```

```
| id             | 57ecb08c130f4632ac606396a66701ef |
| name           | keystone                         |
| type           | identity                         |
+----------------+----------------------------------+
```

**注意**：

此例中的 id 值是随机生成的，可能不尽相同。

### 2. 创建Keystone身份认证服务组件的API endpoint

OpenStack 为每个服务组件都提供了三种 API endpoint，分别是 admin、internal 和 public。admin API endpoint 默认允许修改用户和租户，但是 internal 和 public API endpoint 不能。在生产环境中，出于安全考虑，会将三种 API endpoint 的通信放在独立的网络环境中，public API endpoint 用于互联网中的用户管理虚拟机实例，admin API endpoints 用于内部管理员管理自己的基础设施，internal API endpoint 用于管理 OpenStack 项目中的各个服务组件。OpenStack 项目也支持多个 region，简便起见，本例中默认使用一个 region（RegionOne）。接下来开始创建 API endpoint。

```
# openstack endpoint create --region RegionOne \
  identity public http://controller:5000/v2.0
+--------------+----------------------------------+
| Field        | Value                            |
+--------------+----------------------------------+
| enabled      | True                             |
| id           | 7c0140a19bfb4548bfc086808d14e534 |
| interface    | public                           |
| region       | RegionOne                        |
| region_id    | RegionOne                        |
| service_id   | 57ecb08c130f4632ac606396a66701ef |
| service_name | keystone                         |
| service_type | identity                         |
| url          | http://Controller:5000/v2.0      |
+--------------+----------------------------------+

# openstack endpoint create --region RegionOne \
  identity internal http://controller:5000/v2.0
+--------------+----------------------------------+
| Field        | Value                            |
+--------------+----------------------------------+
| enabled      | True                             |
| id           | bc5bec379c4f4013916df873de7a02d3 |
| interface    | internal                         |
```

```
| region       | RegionOne                        |
| region_id    | RegionOne                        |
| service_id   | 57ecb08c130f4632ac606396a66701ef |
| service_name | keystone                         |
| service_type | identity                         |
| url          | http://Controller:5000/v2.0      |
+--------------+----------------------------------+

#openstack endpoint create --region RegionOne \
  identity admin http://controller:35357/v2.0
+--------------+----------------------------------+
| Field        | Value                            |
+--------------+----------------------------------+
| enabled      | True                             |
| id           | c23227eb685d4d90982f80f6ae48c062 |
| interface    | admin                            |
| region       | RegionOne                        |
| region_id    | RegionOne                        |
| service_id   | 57ecb08c130f4632ac606396a66701ef |
| service_name | keystone                         |
| service_type | identity                         |
| url          | http://Controller:35357/v2.0     |
+--------------+----------------------------------+
```

## 11.3 创建项目、用户和角色

Keystone 身份认证服务组件为 OpenStack 项目中的每个服务组件提供认证服务，这些功能是通过域（domain）、项目（project or tenant）、用户（user）和权限（role）实现的。

 **注意：**

出于简单的目的，本例中使用 default domain。

下文对域、项目、用户和权限均使用英文单词，方便学习。

（1）创建一个管理员权限的 project、user 和 role。

- 创建 admin project。

```
# openstack project create --domain default \
  --description "Admin Project" admin
+-------------+---------------+
| Field       | Value         |
+-------------+---------------+
| description | Admin Project |
| domain_id   | default       |
```

```
| enabled   | True                             |
| id        | efb33dccf25b40b399e92d6351fc7ea9 |
| is_domain | False                            |
| name      | admin                            |
| parent_id | None                             |
+-----------+----------------------------------+
```

**注意:**
此例中的 id 值是随机生成的，可能不尽相同。

- 创建 admin user。

```
# openstack user create --domain default \
  --password-prompt admin
User Password:
Repeat User Password:
+-----------+----------------------------------+
| Field     | Value                            |
+-----------+----------------------------------+
| domain_id | default                          |
| enabled   | True                             |
| id        | fc3820e9803f46308092b409d6323392 |
| name      | admin                            |
+-----------+----------------------------------+
```

**注意:**
务必记录好将要或已经创建的用户及其密码。在后续操作中，会频繁使用这些信息。

- 创建 admin role。

```
# openstack role create admin
+-------+----------------------------------+
| Field | Value                            |
+-------+----------------------------------+
| id    | e089ff227c504e07bda544bc07b0b989 |
| name  | admin                            |
+-------+----------------------------------+
```

- 将 admin role 赋予 admin project 和 admin user。

```
# openstack role add --project admin --user admin admin
```

**注意:**
此命令执行完成后，没有输出。

（2）创建 service project。service project 包含 OpenStack 项目中已添加的服务组件的唯一用户。

```
# openstack project create --domain default \
  --description "Service Project" service
+-------------+----------------------------------+
| Field       | Value                            |
+-------------+----------------------------------+
| description | Service Project                  |
| domain_id   | default                          |
| enabled     | True                             |
| id          | 8af2565380aa40f1828423d892b60a7e |
| is_domain   | False                            |
| name        | service                          |
| parent_id   | None                             |
+-------------+----------------------------------+
```

（3）创建一个非管理员权限 project、user 和 role。

- 创建 demo project。

```
# openstack project create --domain default \
  --description "Demo Project" demo
+-------------+----------------------------------+
| Field       | Value                            |
+-------------+----------------------------------+
| description | Demo Project                     |
| domain_id   | default                          |
| enabled     | True                             |
| id          | 9543b7db999641fda39ff53dc22e7bcb |
| is_domain   | False                            |
| name        | demo                             |
| parent_id   | None                             |
+-------------+----------------------------------+
```

注意：

为 demo project 添加 user 时，不要重复创建 demo project。

- 创建 demo user。

```
# openstack user create --domain default \
  --password-prompt demo
User Password:
Repeat User Password:
```

```
+-----------+------------------------------------+
| Field     | Value                              |
+-----------+------------------------------------+
| domain_id | default                            |
| enabled   | True                               |
| id        | 33cd0cb0e80a4ba6bb5b8c36947f6267   |
| name      | demo                               |
+-----------+------------------------------------+
```

- 创建 user role。

```
# openstack role create user
+-------+----------------------------------+
| Field | Value                            |
+-------+----------------------------------+
| id    | 361c58c26c8c46b49315fb33a3ad5870 |
| name  | user                             |
+-------+----------------------------------+
```

- 将 user role 赋予 demo project 和 demo user。

```
# openstack role add --project demo --user demo user
```

注意：

此命令执行完成后，没有输出。

注意：

重复以上步骤创建更多的 project、user 和 role。

## 11.4 检查配置

在进行下一步操作之前，必须确认之前的操作是否正确。以下操作均在控制节点上运行。

（1）出于安全因素的考虑，关闭以下功能。

编辑/etc/keystone/keystone-paste.ini 文件，在[pipeline:public_api]、[pipeline:admin_api] 和[pipeline:api_v3]三项删除 admin_token_auth。

（2）删除 OS_TOKEN 和 OS_URL 两个变量。

```
# unset OS_TOKEN OS_URL
```

(3）使用 admin 用户申请 authentication token。

```
# openstack --os-auth-url http://controller:35357/v3 \
  --os-project-domain-id default --os-user-domain-id default \
  --os-project-name admin --os-username admin \
  --os-auth-type password token issue
Password:
```

注意：

在 Password 处输入 admin 用户的密码，之前章节已经创建。

(4）使用 demo 用户申请 authentication token。

```
# openstack --os-auth-url http://controller:5000/v3 \
  --os-project-domain-id default --os-user-domain-id default \
  --os-project-name demo --os-username demo --os-auth-type \
  password token issue
Password:
```

注意：

在 Password 处输入 demo 用户的密码，之前章节已经创建。

同时可以观察到，demo（non-admin）用户和 admin（admin）用户在控制节点上所使用的通信端口号是不同的。

另外，在使用 OpenStack 客户端操作命令输入时，需要定义大量的环境变量，操作复杂。11.5 节将解决这个问题。

## 11.5 定义 OpenStack 客户端环境变量脚本

在此之前，在 OpenStack 客户端通过环境变量和操作命令相互组合完成了与 Keystone 身份认证服务组件的通信。为了提高可操作性和工作效率，可以创建一个统一而完整的 openRC 文件，其涵盖了通用变量和特殊变量。

## 11.5.1 创建环境变量脚本

创建 admin project、demo project 和用户的环境变量脚本。

（1）创建文件 admin-openrc.sh 并添加以下内容。

```
export OS_PROJECT_DOMAIN_ID=default
export OS_USER_DOMAIN_ID=default
export OS_PROJECT_NAME=admin
export OS_TENANT_NAME=admin
export OS_USERNAME=admin
export OS_PASSWORD=ADMIN_PASS
export OS_AUTH_URL=http://controller:35357/v3
export OS_IDENTITY_API_VERSION=3
```

使用已经生成的 admin user 密码替换 ADMIN_PASS。

（2）创建文件 demo-openrc.sh 并添加以下内容。

```
export OS_PROJECT_DOMAIN_ID=default
export OS_USER_DOMAIN_ID=default
export OS_PROJECT_NAME=demo
export OS_TENANT_NAME=demo
export OS_USERNAME=demo
export OS_PASSWORD=DEMO_PASS
export OS_AUTH_URL=http://controller:5000/v3
export OS_IDENTITY_API_VERSION=3
```

使用已经生成的 demo user 密码替换 DEMO_PASS。

## 11.5.2 验证

选择已创建完成的环境变量脚本，进行测试。

（1）加载 admin-openrc.sh 文件，自动生成环境变量。

```
# source admin-openrc.sh
```

(2) 请求 authentication token。

```
# openstack token issue
+------------+----------------------------------+
| Field      | Value                            |
+------------+----------------------------------+
| expires    | 2016-07-18T04:18:42.529743Z      |
| id         | 249d1e3ee78b4f888d759f40be8acb2f |
| project_id | efb33dccf25b40b399e92d6351fc7ea9 |
| user_id    | fc3820e9803f46308092b409d6323392 |
+------------+----------------------------------+
```

与 11.4 节的操作相比，操作简单，且不易出错。

# 第 12 章 镜像（Glance）服务安装配置

Glance 镜像服务组件对 OpenStack 项目很重要，它接收各种 API 请求，包括磁盘或服务器镜像、用户的元数据定义和 Nova 计算服务组件调用等。对于保存镜像的存储，存储形式有很多，其中就包括 Swift 对象存储。

Glance 镜像服务组件中运行很多周期性的进程支持缓存，副本服务保证了数据的一致性和可用性。

## 12.1 安装和配置

本节详细描述如何在控制节点上安装和配置 Glance 镜像服务组件。简单起见，我们使用普通文件系统作为存储镜像的介质。

### 12.1.1 准备

在安装和配置 Glance 镜像服务组件之前，首先创建数据库、服务证书（service credential）和 API endpoint。

(1) 创建数据库。

① 在操作系统终端连接数据库。

```
# mysql -u root -p
```

因为我们使用 root 用户登录,所以在提示输入密码时,按回车键即可建立连接。

② 创建 Glance 数据库。

```
MariaDB [(none)]> CREATE DATABASE glance;
```

③ Glance 数据库的访问权限设置。

```
GRANT ALL PRIVILEGES ON glance.* TO 'glance'@'localhost' \
  IDENTIFIED BY 'GLANCE_DBPASS';
GRANT ALL PRIVILEGES ON glance.* TO 'glance'@'%' \
  IDENTIFIED BY 'GLANCE_DBPASS';
```

替换 GLANCE_DBPASS 为合适的密码。

④ 退出数据库。

(2) 加载 admin user 的环境变量。

```
# source admin-openrc.sh
```

(3) 创建 Identity 服务凭据。

① 创建 Glance 用户。

```
# openstack user create --domain default --password-prompt glance
User Password:
Repeat User Password:
+-----------+----------------------------------+
| Field     | Value                            |
+-----------+----------------------------------+
| domain_id | default                          |
| enabled   | True                             |
| id        | 2c27875abf014b33949cf5de9ebdf578 |
| name      | glance                           |
+-----------+----------------------------------+
```

② 将 admin role 赋予 glance user 和 service project。

```
# openstack role add --project service --user glance admin
```

> **注意:**
> 此命令执行完成后,没有输出。

③ 创建 Glance 镜像的 service entity。

```
# openstack service create --name glance \
  --description "OpenStack Image service" image
+-------------+----------------------------------+
| Field       | Value                            |
+-------------+----------------------------------+
| description | OpenStack Image service          |
| enabled     | True                             |
| id          | bccb49da242d4e3f806f0ca563c19584 |
| name        | glance                           |
| type        | image                            |
+-------------+----------------------------------+
```

(4) 创建 Glance 镜像服务组件的 API endpoint。

```
# openstack endpoint create --region RegionOne \
  image public http://Controller:9292
+--------------+----------------------------------+
| Field        | Value                            |
+--------------+----------------------------------+
| enabled      | True                             |
| id           | db66e030c61144f4a02dff90612dca7b |
| interface    | public                           |
| region       | RegionOne                        |
| region_id    | RegionOne                        |
| service_id   | bccb49da242d4e3f806f0ca563c19584 |
| service_name | glance                           |
| service_type | image                            |
| url          | http://Controller:9292           |
+--------------+----------------------------------+

# openstack endpoint create --region RegionOne \
  image internal http://Controller:9292
+--------------+----------------------------------+
| Field        | Value                            |
+--------------+----------------------------------+
| enabled      | True                             |
| id           | 8453e2db71ce47cc8ec91f7275fbe464 |
| interface    | internal                         |
| region       | RegionOne                        |
| region_id    | RegionOne                        |
| service_id   | bccb49da242d4e3f806f0ca563c19584 |
| service_name | glance                           |
| service_type | image                            |
| url          | http://Controller:9292           |
+--------------+----------------------------------+

# openstack endpoint create --region RegionOne \
  image admin http://Controller:9292
```

```
+--------------+------------------------------------+
| Field        | Value                              |
+--------------+------------------------------------+
| enabled      | True                               |
| id           | 6d42fac1296d47dda665f4ec36b105c1   |
| interface    | admin                              |
| region       | RegionOne                          |
| region_id    | RegionOne                          |
| service_id   | bccb49da242d4e3f806f0ca563c19584   |
| service_name | glance                             |
| service_type | image                              |
| url          | http://Controller:9292             |
+--------------+------------------------------------+
```

## 12.1.2 安装和配置 Glance 镜像服务组件

下面开始安装和配置 Glance 镜像服务组件，由于版本不同，可能一些配置文件的内容也不同，需要修改或者添加一些内容。省略号（…）代表配置文件中需要保留的默认配置信息。

（1）安装软件包。

```
# zypper install openstack-glance python-glanceclient
```

（2）编辑文件/etc/glance/glance-api.conf，完成以下操作。

① 在[database]项，配置数据库连接。

```
[database]
...
connection = mysql+pymysql://glance:GLANCE_DBPASS@Controller/glance
```

使用已经定义的 Glance 数据库密码替换 GLANCE_DBPASS。

② 在[keystone_authtoken]和[paste_deploy]项，配置 Keystone 身份认证服务组件访问。

```
[keystone_authtoken]
...
auth_uri = http://controller:5000
auth_url = http://controller:35357
auth_plugin = password
project_domain_id = default
user_domain_id = default
project_name = service
username = glance
```

```
password = GLANCE_PASS

[paste_deploy]
...
flavor = keystone
```

使用已经定义的 Glance 用户密码替换 GLANCE_PASS。

 **注意:**

注释或者删除 [keystone_authtoken]项中的其他内容。

③ 在[glance_store]项，配置虚拟机镜像的存储形式和存储路径。

```
[glance_store]
...
default_store = file
filesystem_store_datadir = /var/lib/glance/images/
```

④ 在[DEFAULT]项，关闭 notification_driver 功能，设置值为 noop。配置 Ceilometer 计量服务组件需要打开该功能。

```
[DEFAULT]
...
notification_driver = noop
```

⑤ 打开日志记录功能，方便问题跟踪和解决。

```
[DEFAULT]
...
verbose = True
```

(3) 编辑文件/etc/glance/glance-registry.conf，完成以下操作。

① 在[database]项，配置数据库连接。

```
[database]
...
connection = mysql+pymysql://glance:GLANCE_DBPASS@Controller/glance
```

使用已经定义的 Glance 数据库密码替换 GLANCE_DBPASS。

② 在[keystone_authtoken]和[paste_deploy]项，配置 Keystone 身份认证服务组件连接。

```
[keystone_authtoken]
...
auth_uri = http://controller:5000
auth_url = http://controller:35357
auth_plugin = password
project_domain_id = default
user_domain_id = default
project_name = service
username = glance
password = GLANCE_PASS

[paste_deploy]
...
flavor = keystone
```

使用已经定义的 Glance 用户密码替换 GLANCE_PASS。

  注意：

> 注释或者删除 [keystone_authtoken] 项中的其他内容。

③ 在[DEFAULT]项，关闭 notification_driver 功能，设置值为 noop。配置 Ceilometer 计量服务组件需要打开该功能。

```
[DEFAULT]
...
notification_driver = noop
```

④ 打开日志记录功能，方便问题跟踪和解决。

```
[DEFAULT]
...
verbose = True
```

## 12.1.3 安装完成

启动 Glance 镜像服务组件并设置为开机自启动。

```
# systemctl enable openstack-glance-api.service \
  openstack-glance-registry.service
# systemctl start openstack-glance-api.service \
  openstack-glance-registry.service
```

## 12.2 验证

下面使用一个简单的 Linux 镜像文件 CirrOS 测试配置完成的 Glance 镜像服务组件。以下操作均在控制节点上进行。

(1) 在操作系统终端执行以下命令，在 admin-openrc.sh 和 demo-openrc.sh 文件中增加环境变量值。

```
# echo "export OS_IMAGE_API_VERSION=2" \
  | tee -a admin-openrc.sh demo-openrc.sh
```

(2) 加载 admin-openrc.sh 文件，自动生成环境变量。

```
# source admin-openrc.sh
```

(3) 下载 CirrOS 镜像。

```
# wget \
http://download.cirros-cloud.net/0.3.4/cirros-0.3.4-x86_64-disk.img
```

注意：

如果 wget 不能使用，则可能是因为 wget 软件包没有安装。

可以使用 rpm -qa|grep -I wget 命令检查 wget 软件包是否安装。

(4) 上传 CirrOS 镜像文件到 Glance 镜像服务组件中。

```
# glance image-create --name "cirros" \
  --file cirros-0.3.4-x86_64-disk.img \
  --disk-format qcow2 --container-format bare \
  --visibility public --progress
```

```
+------------------+--------------------------------------+
| Property         | Value                                |
+------------------+--------------------------------------+
| checksum         | ee1eca47dc88f4879d8a229cc70a07c6     |
| container_format | bare                                 |
| created_at       | 2016-02-23T15:04:07Z                 |
| disk_format      | qcow2                                |
| id               | 0ee987b9-4a8e-4908-8ab5-833b781f0502 |
| min_disk         | 0                                    |
| min_ram          | 0                                    |
| name             | cirros                               |
| owner            | efb33dccf25b40b399e92d6351fc7ea9     |
| protected        | False                                |
| size             | 13287936                             |
| status           | active                               |
| tags             | []                                   |
| updated_at       | 2016-02-23T15:04:08Z                 |
| virtual_size     | None                                 |
| visibility       | public                               |
+------------------+--------------------------------------+
```

**注意：**

此例中的 id 值是随机生成的，可能不尽相同。

(5) 确认镜像。

```
# glance image-list
+--------------------------------------+--------+
| ID                                   | Name   |
+--------------------------------------+--------+
| 0ee987b9-4a8e-4908-8ab5-833b781f0502 | cirros |
+--------------------------------------+--------+
```

# 第 13 章 计算（Nova）服务安装配置

OpenStack 项目中的 Nova 计算服务是 Infrastructure as a Service（IaaS）云计算平台的核心服务组件，控制着虚拟机实例和网络功能。通过对用户和项目的设置，管理对 OpenStack 云资源的访问。

Nova 计算服务组件没有创造新的虚拟化技术（如 KVM 或 Xen 等虚拟化技术），而是定义了与底层虚拟化技术进行交互的各种驱动，从而实现更多的功能，通过基于 Web 的 API 将这些功能公开。

## 13.1 安装和配置（控制节点）

本节详细描述如何在控制节点上安装和配置 Nova 计算服务组件。

### 13.1.1 准备

在安装和配置 Nova 计算服务组件之前，首先创建数据库、服务证书（service credential）和 API endpoint。

(1) 创建数据库。

① 在操作系统终端连接数据库。

```
# mysql -u root -p
```

因为我们使用 root 用户登录，所以在提示输入密码时，按回车键即可自动建立连接。

② 创建 Nova 数据库。

```
MariaDB [(none)]> CREATE DATABASE nova;
```

③ Nova 数据库的访问权限设置。

```
GRANT ALL PRIVILEGES ON nova.* TO 'nova'@'localhost' \
  IDENTIFIED BY 'NOVA_DBPASS';
GRANT ALL PRIVILEGES ON nova.* TO 'nova'@'%' \
  IDENTIFIED BY 'NOVA_DBPASS';
```

替换 NOVA_DBPASS 为合适的密码。

④ 退出数据库。

(2) 加载 admin 用户的环境变量。

```
# source admin-openrc.sh
```

(3) 创建认证服务凭据。

① 创建 Nova 用户。

```
# openstack user create --domain default --password-prompt nova
User Password:
Repeat User Password:
+-----------+----------------------------------+
| Field     | Value                            |
+-----------+----------------------------------+
| domain_id | default                          |
| enabled   | True                             |
| id        | 11839df86ee94ca7a36ee514a4eaf316 |
| name      | nova                             |
+-----------+----------------------------------+
```

② 将 admin role 赋予 glance user 和 service project。

```
# openstack role add --project service --user nova admin
```

**注意:**

此命令执行完成后，没有输出。

③ 创建 Nova 计算服务 service entity。

```
# openstack service create --name nova \
  --description "OpenStack Nova service" image
+-------------+----------------------------------+
| Field       | Value                            |
+-------------+----------------------------------+
| description | OpenStack Compute                |
| enabled     | True                             |
| id          | f407314c45df4f8faa04bb331aa8cddd |
| name        | nova                             |
| type        | compute                          |
+-------------+----------------------------------+
```

(4) 创建 Nova 计算服务组件的 API endpoint。

```
# openstack endpoint create --region RegionOne \
  compute public http://Controller:8774/v2/%\(tenant_id\)s
+--------------+-------------------------------------------+
| Field        | Value                                     |
+--------------+-------------------------------------------+
| enabled      | True                                      |
| id           | fd71e27d613348fe89accdcd89c8453b          |
| interface    | public                                    |
| region       | RegionOne                                 |
| region_id    | RegionOne                                 |
| service_id   | f407314c45df4f8faa04bb331aa8cddd          |
| service_name | nova                                      |
| service_type | compute                                   |
| url          | http://Controller:8774/v2/%(tenant_id)s   |
+--------------+-------------------------------------------+

# openstack endpoint create --region RegionOne \
  compute internal http://Controller:8774/v2/%\(tenant_id\)s
+--------------+-------------------------------------------+
| Field        | Value                                     |
+--------------+-------------------------------------------+
| enabled      | True                                      |
| id           | 05a70eede5544526bfb8ac2070f525bf          |
| interface    | internal                                  |
| region       | RegionOne                                 |
| region_id    | RegionOne                                 |
| service_id   | f407314c45df4f8faa04bb331aa8cddd          |
| service_name | nova                                      |
| service_type | compute                                   |
| url          | http://Controller:8774/v2/%(tenant_id)s   |
+--------------+-------------------------------------------+
```

```
# openstack endpoint create --region RegionOne \
  compute admin http://Controller:8774/v2/%\(tenant_id\)s
+--------------+----------------------------------------------+
| Field        | Value                                        |
+--------------+----------------------------------------------+
| enabled      | True                                         |
| id           | 3908be21333f4d64951f0a884bf756fb             |
| interface    | admin                                        |
| region       | RegionOne                                    |
| region_id    | RegionOne                                    |
| service_id   | f407314c45df4f8faa04bb331aa8cddd             |
| service_name | nova                                         |
| service_type | compute                                      |
| url          | http://Controller:8774/v2/%(tenant_id)s      |
+--------------+----------------------------------------------+
```

## 13.1.2 安装和配置 Nova 计算服务组件

下面开始安装和配置 Nova 计算服务组件，由于版本的不同，可能一些配置文件的内容也不同，需要修改或者添加一些内容。省略号（...）代表配置文件中需要保留的默认配置信息。

（1）安装软件包。

```
# zypper install openstack-nova-api openstack-nova-scheduler \
  openstack-nova-cert openstack-nova-conductor \
  openstack-nova-consoleauth openstack-nova-novncproxy \
  python-novaclient iptables
```

（2）编辑文件/etc/nova/nova.conf，完成以下操作。

① 在[database]项，配置数据库连接。

```
[database]
...
connection = mysql+pymysql://nova:NOVA_DBPASS@controller/nova
```

使用已经定义的 Nova 数据库密码替换 NOVA_DBPASS。

② 在[DEFAULT]和[oslo_messaging_rabbit]项，配置 RabbitMQ 消息队列连接。

```
[DEFAULT]
...
rpc_backend = rabbit
```

```
[oslo_messaging_rabbit]
...
rabbit_host = Controller
rabbit_userid = openstack
rabbit_password = RABBIT_PASS
```

使用已经在 RabbitMQ 中定义的 OpenStack 用户密码替换 RABBIT_PASS。

③ 在[DEFAULT]和[keystone_authtoken]项，配置 Keystone 身份认证服务组件访问。

```
[DEFAULT]
...
auth_strategy = keystone

[keystone_authtoken]
...
auth_uri = http://controller:5000
auth_url = http://controller:35357
auth_plugin = password
project_domain_id = default
user_domain_id = default
project_name = service
username = nova
password = NOVA_PASS
```

使用已经定义的 Nova 用户密码替换 NOVA_PASS。

**注意：**

注释或者删除 [keystone_authtoken] 项中的其他内容。

④ 在[DEFAULT]项，配置 my_ip 参数为控制节点的管理/数据网络 IP 地址。

```
[DEFAULT]
...
my_ip = 172.168.1.11
```

⑤ 在[DEFAULT]项，定义 Nova 支持 Neutron 网络服务组件。

```
[DEFAULT]
...
network_api_class = nova.network.neutronv2.api.API
```

```
    security_group_api = neutron
    linuxnet_interface_driver = nova.network.linux_net.NeutronLinuxBridge
InterfaceDriver
    firewall_driver = nova.virt.firewall.NoopFirewallDriver
```

**注意：**
在默认情况下，Nova 计算服务组件使用自己的防火墙功能，但是 Neutron 网络服务组件也有自己的防火墙，为避免冲突和规则重复设置，需要关闭 Nova 计算服务组件的防火墙功能。

⑥ 在[vnc]项，配置 VNC 使用控制节点的管理/数据网络 IP 地址。

```
[vnc]
...
vncserver_listen = $my_ip
vncserver_proxyclient_address = $my_ip
```

⑦ 在[glance]项，定义 Glance 镜像服务组件所在的主机节点。

```
[glance]
...
host = controller
```

⑧ 在[oslo_concurrency]项，配置 lock 路径。

```
[oslo_concurrency]
...
lock_path = /var/run/nova
```

⑨ 在[DEFAULT]项，停止使用 EC2 API。

```
[DEFAULT]
...
enabled_apis=osapi_compute,metadata
```

⑩ 打开日志记录功能，方便问题跟踪和解决。

```
[DEFAULT]
...
verbose = True
```

## 13.1.3 安装完成

启动 Nova 计算服务组件并设置为开机自启动。

```
# systemctl enable openstack-nova-api.service \
  openstack-nova-cert.service openstack-nova-consoleauth.service \
  openstack-nova-scheduler.service \
  openstack-nova-conductor.service \
  openstack-nova-novncproxy.service
# systemctl start openstack-nova-api.service \
  openstack-nova-cert.service openstack-nova-consoleauth.service \
  openstack-nova-scheduler.service \
  openstack-nova-conductor.service \
  openstack-nova-novncproxy.service
```

## 13.2 安装和配置（计算节点）

本节详细描述如何在计算节点上安装和配置 Nova 计算服务组件。Nova 计算服务组件支持多种虚拟化技术，如 KVM、Xen 和 QEMU 等。Nova 计算服务组件支持横向扩展，可以利用自动化工具以同样的方式安装和配置多个计算节点，每个计算节点都需要一个唯一的 IP 地址。本节采用 QEMU 虚拟化技术进行配置。

### 13.2.1 安装和配置 Nova 计算服务组件

下面开始安装和配置 nova 计算服务组件，由于版本的不同，可能一些配置文件的内容也不同，需要修改或者添加一些内容。省略号（...）代表配置文件中需要保留的默认配置信息。

（1）安装软件包。

```
# zypper install openstack-nova-compute genisoimage kvm libvirt
```

（2）编辑文件/etc/nova/nova.conf，完成以下操作。

① 在[DEFAULT]和[oslo_messaging_rabbit]项，配置 RabbitMQ 消息队列连接。

```
[DEFAULT]
...
rpc_backend = rabbit

[oslo_messaging_rabbit]
...
rabbit_host = controller
rabbit_userid = openstack
rabbit_password = RABBIT_PASS
```

使用已经在 RabbitMQ 中定义的 OpenStack 用户密码替换 RABBIT_PASS。

② 在[DEFAULT]和[keystone_authtoken]项，配置 Keystone 身份认证服务组件连接。

```
[DEFAULT]
...
auth_strategy = keystone

[keystone_authtoken]
...
auth_uri = http://controller:5000
auth_url = http://controller:35357
auth_plugin = password
project_domain_id = default
user_domain_id = default
project_name = service
username = nova
password = NOVA_PASS
```

使用已经定义的 Nova 用户密码替换 NOVA_PASS。

 注意：

注释或者删除 [keystone_authtoken] 项中的其他内容。

③ 在[DEFAULT]项，配置 my_ip 参数为计算节点的管理/数据网络 IP 地址。

```
[DEFAULT]
...
my_ip = MANAGEMENT_INTERFACE_IP_ADDRESS
```

如果有多个计算节点，则以此类推。

④ 在[DEFAULT]项，定义 Nova 支持 Neutron 网络服务组件。

```
[DEFAULT]
...
network_api_class = nova.network.neutronv2.api.API
security_group_api = neutron
linuxnet_interface_driver = nova.network.linux_net.NeutronLinuxBridge
InterfaceDriver
firewall_driver = nova.virt.firewall.NoopFirewallDriver
```

⑤ 在[vnc]项，定义和配置远程控制台访问。

```
[vnc]
...
enabled = True
vncserver_listen = 0.0.0.0
vncserver_proxyclient_address = $my_ip
novncproxy_base_url = http://controller:6080/vnc_auto.html
```

注意：

novncproxy_base_url 参数定义的 Controller 必须可以被解析，如果不能被解析，则需要在/etc/hosts 中配置名字解析或者用 IP 地址代替 Controller。

⑥ 在[glance]项，配置 Glance 镜像服务组件所在的节点位置。

```
[glance]
...
host = controller
```

⑦ 在[oslo_concurrency]项，配置 lock 路径。

```
[oslo_concurrency]
...
lock_path = /var/run/nova
```

⑧ 打开日志记录功能，方便问题跟踪和解决。

```
[DEFAULT]
```

```
...
verbose = True
```

（3）加载 nbd 内核模块。

```
# modprobe nbd
```

（4）编辑文件/etc/modules-load.d/nbd.conf，确保每次操作系统重启时，能够自动加载 nbd 内核模块。

（5）检查 Compute Node 是否支持硬件加速（hardware acceleration）功能。

```
# egrep -c '(vmx|svm)' /proc/cpuinfo
```

如果返回值为 1 或者更大数值，则证明计算节点支持硬件加速功能。

如果返回值为 0，则证明计算节点不支持硬件加速功能。

编辑文件/etc/nova/nova.conf 的[libvirt]项，修改 virt_type 值为 qemu。

```
[libvirt]
...
virt_type = qemu
```

## 13.2.2 安装完成

启动 Nova 计算服务组件并设置为开机自启动。

```
# systemctl enable libvirtd.service openstack-nova-compute.service
# systemctl start libvirtd.service openstack-nova-compute.service
```

## 13.3 验证

下面检查一下配置完成的 Nova 计算服务组件。所有的操作均在控制节点上进行。

（1）加载 admin-openrc.sh 文件，自动生成环境变量。

```
# source admin-openrc.sh
```

(2) 确认 Nova 计算服务组件是否已经成功运行和注册。

```
# nova service-list
+----+------------------+------------+----------+---------+-------+----------------------------+-----------------+
| Id | Binary           | Host       | Zone     | Status  | State | Updated_at                 | Disabled Reason |
+----+------------------+------------+----------+---------+-------+----------------------------+-----------------+
| 1  | nova-scheduler   | Controller | internal | enabled | up    | 2016-04-12T03:53:27.000000 | -               |
| 2  | nova-conductor   | Controller | internal | enabled | up    | 2016-04-12T03:53:20.000000 | -               |
| 3  | nova-consoleauth | Controller | internal | enabled | up    | 2016-04-12T03:53:27.000000 | -               |
| 4  | nova-cert        | Controller | internal | enabled | up    | 2016-04-12T03:53:18.000000 | -               |
| 5  | nova-compute     | Compute01  | nova     | enabled | up    | 2016-04-12T03:53:19.000000 | -               |
| 6  | nova-compute     | Compute02  | nova     | enabled | up    | 2016-04-12T03:53:23.000000 | -               |
+----+------------------+------------+----------+---------+-------+----------------------------+-----------------+
```

(3) 确认与 Keystone 身份认证服务组件 API endpoint 的连接。

```
# nova endpoints
+-----------+-------------------------------------------------------------+
| nova      | Value                                                       |
+-----------+-------------------------------------------------------------+
| id        | 05a70eede5544526bfb8ac2070f525bf                            |
| interface | internal                                                    |
| region    | RegionOne                                                   |
| region_id | RegionOne                                                   |
| url       | http://Controller:8774/v2/efb33dccf25b40b399e92d6351fc7ea9  |
+-----------+-------------------------------------------------------------+
| nova      | Value                                                       |
+-----------+-------------------------------------------------------------+
| id        | 3908be21333f4d64951f0a884bf756fb                            |
| interface | admin                                                       |
| region    | RegionOne                                                   |
| region_id | RegionOne                                                   |
| url       | http://Controller:8774/v2/efb33dccf25b40b399e92d6351fc7ea9  |
+-----------+-------------------------------------------------------------+
| nova      | Value                                                       |
+-----------+-------------------------------------------------------------+
| id        | fd71e27d613348fe89accdcd89c8453b                            |
| interface | public                                                      |
| region    | RegionOne                                                   |
| region_id | RegionOne                                                   |
| url       | http://Controller:8774/v2/efb33dccf25b40b399e92d6351fc7ea9  |
+-----------+-------------------------------------------------------------+
```

> **注意：**
>
> 执行此命令时输出的 warning 信息可以忽略。
>
> 执行此命令时会输出已经安装的所有服务组件的 endpoint，在此只列举 Nova 部分，其他服务组件没有一一列出。

(4) 确认与 Glance 镜像服务组件的连接。

```
# nova image-list
+--------------------------------------+--------+--------+--------+
| ID                                   | Name   | Status | Server |
+--------------------------------------+--------+--------+--------+
| 0ee987b9-4a8e-4908-8ab5-833b781f0502 | cirros | ACTIVE |        |
+--------------------------------------+--------+--------+--------+
```

# 第 14 章 网络（Neutron）服务安装配置

本章将介绍如何安装和配置 Neutron 网络服务组件。Neutron 网络服务组件通过提供 API 使用户可以定义网络连接类型和 IP 网络地址、基于三层路由转发和 NAT 的负载均衡、防火墙和 IPSec VPN 等，提供了多种网络技术，丰富了网络功能，驱动了 OpenStack 云计算网络的建设。

注意：

本章使用 Neutron 网络服务组件进行网络模块的部署，务必牢记不要同时在计算节点上启用 nova-network 服务。二者具有相同的功能，同时启用会产生冲突，所以只能部署其中一种。

## 14.1 安装和配置（控制节点）

本节详细描述如何在控制节点上安装和配置 Neutron 网络服务组件。

## 14.1.1 准备

在安装和配置 Neutron 网络服务组件之前，首先创建数据库、服务证书（service credential）和 API endpoint。

(1) 创建数据库。

① 在操作系统终端连接数据库。

```
# mysql -u root -p
```

因为我们使用 root 用户登录，所以在提示输入密码时，按回车键即可建立连接。

② 创建 Neutron 数据库。

```
MariaDB [(none)]> CREATE DATABASE neutron;
```

③ Neutron 数据库的访问权限设置。

```
GRANT ALL PRIVILEGES ON neutron.* TO 'neutron'@'localhost' \
  IDENTIFIED BY 'NEUTRON_DBPASS';
GRANT ALL PRIVILEGES ON neutron.* TO 'neutron'@'%' \
  IDENTIFIED BY 'NEUTRON_DBPASS';
```

替换 NEUTRON_DBPASS 为合适的密码。

④ 退出数据库。

(2) 加载 admin user 的环境变量。

```
# source admin-openrc.sh
```

(3) 创建 Identity 服务凭据。

① 创建 Neutron 用户。

```
# openstack user create --domain default --password-prompt neutron
User Password:
Repeat User Password:
```

```
+------------+------------------------------------+
| Field      | Value                              |
+------------+------------------------------------+
| domain_id  | default                            |
| enabled    | True                               |
| id         | 09aa4d42b11648f29845d4b5e0120283   |
| name       | neutron                            |
+------------+------------------------------------+
```

② 将 admin role 赋予 neutron user 和 service project。

```
# openstack role add --project service --user neutron admin
```

 注意：

此命令执行完成后，没有输出。

③ 创建 Neutron 网络服务 service entity。

```
# openstack service create --name neutron \
  --description "OpenStack Neutron service" image
+-------------+----------------------------------+
| Field       | Value                            |
+-------------+----------------------------------+
| description | OpenStack Compute                |
| enabled     | True                             |
| id          | f407314c45df4f8faa04bb331aa8cddd |
| name        | nova                             |
| type        | compute                          |
+-------------+----------------------------------+
```

（4）创建 Neutron 网络服务组件的 API endpoint。

```
# openstack endpoint create --region RegionOne \
  network public http://controller:9696
+--------------+----------------------------------+
| Field        | Value                            |
+--------------+----------------------------------+
| enabled      | True                             |
| id           | 9f5fad1ac4de4cf6904c715228ce364f |
| interface    | public                           |
| region       | RegionOne                        |
| region_id    | RegionOne                        |
| service_id   | 3ee88be861b54e1cac40fef27c86751e |
| service_name | neutron                          |
| service_type | network                          |
| url          | http://Controller:9696           |
+--------------+----------------------------------+

# openstack endpoint create --region RegionOne \
  network internal http://controller:9696
```

```
+--------------+---------------------------------+
| Field        | Value                           |
+--------------+---------------------------------+
| enabled      | True                            |
| id           | 9a07527d1e2044c9ae801ffd7116e383 |
| interface    | internal                        |
| region       | RegionOne                       |
| region_id    | RegionOne                       |
| service_id   | 3ee88be861b54e1cac40fef27c86751e |
| service_name | neutron                         |
| service_type | network                         |
| url          | http://Controller:9696          |
+--------------+---------------------------------+

# openstack endpoint create --region RegionOne \
  network admin http://controller:9696
+--------------+---------------------------------+
| Field        | Value                           |
+--------------+---------------------------------+
| enabled      | True                            |
| id           | 4af00cc9caab4b96834180c05841acc0 |
| interface    | admin                           |
| region       | RegionOne                       |
| region_id    | RegionOne                       |
| service_id   | 3ee88be861b54e1cac40fef27c86751e |
| service_name | neutron                         |
| service_type | network                         |
| url          | http://Controller:9696          |
+--------------+---------------------------------+
```

## 14.1.2 配置 Neutron 网络服务组件

本章介绍两种关于 Neutron 网络服务组件的设计架构。

- 架构一比较简单，虚拟机实例连接公共网络（Public Network），没有自服务网络（Self-Service Network）、路由和浮动 IP 地址（Floating IP Address），只有管理员（admin）或者有权限的用户可以操作 Public（Provider）Network。

- 架构二是在架构一的基础上增加 Layer 3 服务，支持自服务 Self-Service（Private）网络。demo 用户和有权限的用户都可以管理自服务网络（Self-Service Network），在自服务网络（Self-Service Network）和 Provider 网络间路由信息，进行通信。同时，分配给虚拟机实例的 Floating IP 通过自服务网络（Self-Service Network）可以访问外部网络（External Network，如互联网）。

注意：
架构二也支持虚拟机实例使用 Public（Provider）Network。

## 1. 架构一配置 (Provider Network)

以下操作均在控制节点上完成。

(1) 安装 Networking 组件。

```
# zypper install --no-recommends openstack-neutron \
  openstack-neutron-server openstack-neutron-linuxbridge-agent \
  openstack-neutron-dhcp-agent openstack-neutron-metadata-agent \
  ipset
```

(2) 编辑文件/etc/neutron/neutron.conf，完成以下操作。

① 在[database]项，配置数据库连接。

```
[database]
...
connection = mysql+pymysql://neutron:NEUTRON_DBPASS@controller/neutron
```

使用已经定义的 Neutron 数据库密码替换 NEUTRON_DBPASS。

② 在[DEFAULT]项，启用 Layer 2 (ML2) plug-in 和禁用 service_plugins。

```
[DEFAULT]
...
core_plugin = ml2
service_plugins =
```

③ 在[DEFAULT]和[oslo_messaging_rabbit]项，配置 RabbitMQ 消息队列连接。

```
[DEFAULT]
...
rpc_backend = rabbit

[oslo_messaging_rabbit]
...
rabbit_host = controller
rabbit_userid = openstack
rabbit_password = RABBIT_PASS
```

使用已经在 RabbitMQ 中定义的 OpenStack 用户密码替换 RABBIT_PASS。

④ 在[DEFAULT]和[keystone_authtoken]项，配置 Keystone 身份认证服务组件访问。

```
[DEFAULT]
...
auth_strategy = keystone

[keystone_authtoken]
...
auth_uri = http://controller:5000
auth_url = http://controller:35357
auth_plugin = password
project_domain_id = default
user_domain_id = default
project_name = service
username = neutron
password = NEUTRON_PASS
```

使用已经定义的 Neutron 用户密码替换 NEUTRON_PASS。

 注意：

注释或者删除[keystone_authtoken]项中的其他内容。

⑤ 在[DEFAULT]和[nova]项，配置 Networking 通知 Compute 关于网络拓扑的改变。

```
[DEFAULT]
...
notify_nova_on_port_status_changes = True
notify_nova_on_port_data_changes = True
nova_url = http://controller:8774/v2

[nova]
...
auth_url = http://controller:35357
auth_plugin = password
project_domain_id = default
user_domain_id = default
region_name = RegionOne
project_name = service
```

```
username = nova
password = NOVA_PASS
```

使用已经定义的 Nova 用户密码替换 NOVA_PASS。

⑥ 打开日志记录功能，方便问题跟踪和解决。

```
[DEFAULT]
...
verbose = True
```

（3）配置 Layer 2 (ML2) plug-in 模块。

ML2 plug-in 使用 bridge 机制为虚拟机实例桥接和交换网络信息。

① 编辑文件/etc/neutron/plugins/ml2/ml2_conf.ini，在[ml2]项，完成以下操作。

```
[ml2]
...
type_drivers = flat,vlan
tenant_network_types =
mechanism_drivers = linuxbridge
extension_drivers = port_security
```

 错误：

配置完成后，切勿删除 type_drivers 的值。

② 编辑文件/etc/neutron/plugins/ml2/ml2_conf.ini，在[ml2_type_flat]项，完成以下操作。

```
[ml2_type_flat]
...
flat_networks = public
```

③ 编辑文件/etc/neutron/plugins/ml2/ml2_conf.ini，在[securitygroup]项，启用 ipset 提高 Security Group 规则的效率。

```
[securitygroup]
...
enable_ipset = True
```

(4) 配置 bridge agent。

bridge agent 为虚拟机实例建立了 Layer 2（桥接和交换网络信息）虚拟网络，提供了实现 Private Network 需要的 VXLAN 和处理 Security Group。

① 编辑/etc/neutron/plugins/ml2/linuxbridge_agent.ini 文件，在[linux_bridge]项，将 Public Network 虚拟网络端口映射到物理网络端口。

```
[linux_bridge]
physical_interface_mappings = public:PUBLIC_INTERFACE_NAME
```

使用已经定义的物理网络端口替换 PUBLIC_INTERFACE_NAME，针对此例，应该是 eth1。

② 编辑文件/etc/neutron/plugins/ml2/linuxbridge_agent.ini，在[vxlan]项，禁用 VXLAN 功能。

```
[vxlan]
enable_vxlan = False
```

③ 编辑文件 /etc/neutron/plugins/ml2/linuxbridge_agent.ini，在[agent]项，启用 ARP 欺骗保护。

```
[agent]
...
prevent_arp_spoofing = True
```

④ 编辑文件 /etc/neutron/plugins/ml2/linuxbridge_agent.ini，在[securitygroup]项，启用 Security Group 和配置 iptables firewall driver。

```
[securitygroup]
...
enable_security_group = True
firewall_driver                                                    =
neutron.agent.linux.iptables_firewall.IptablesFirewallDriver
```

(5) 配置 DHCP agent。

DHCP agent 为虚拟网络提供了 DHCP 服务，也可以说为虚拟机实例提供了 IP 地址。

① 编辑/etc/neutron/dhcp_agent.ini 文件，在[DEFAULT]项，配置 Linux bridge interface driver 和 Dnsmasq DHCP driver，启用 metadata 分离以使虚拟机实例访问互联网上的 metadata。

```
[DEFAULT]
...
interface_driver = neutron.agent.linux.interface.BridgeInterfaceDriver
dhcp_driver = neutron.agent.linux.dhcp.Dnsmasq
enable_isolated_metadata = True
```

② 打开日志记录功能，方便问题跟踪和解决。

```
[DEFAULT]
...
verbose = True
```

## 2．架构二配置（Self-Service Network）

以下操作均在控制节点上完成。

（1）安装 Networking 组件。

```
# zypper install --no-recommends openstack-neutron \
  openstack-neutron-server openstack-neutron-linuxbridge-agent \
  openstack-neutron-l3-agent openstack-neutron-dhcp-agent \
  openstack-neutron-metadata-agent ipset
```

（2）编辑文件/etc/neutron/neutron.conf，完成以下操作。

① 在[database]项，配置数据库连接。

```
[database]
...
connection = mysql+pymysql://neutron:NEUTRON_DBPASS@controller/neutron
```

使用已经定义的 Neutron 数据库密码替换 NEUTRON_DBPASS。

② 在[DEFAULT]项，启用 Layer 2 (ML2) plug-in、路由服务和 IP 地址重叠功能。

```
[DEFAULT]
...
```

```
core_plugin = ml2
service_plugins = router
allow_overlapping_ips = True
```

③ 在[DEFAULT]和[oslo_messaging_rabbit]项，配置 RabbitMQ 消息队列连接。

```
[DEFAULT]
...
rpc_backend = rabbit

[oslo_messaging_rabbit]
...
rabbit_host = controller
rabbit_userid = openstack
rabbit_password = RABBIT_PASS
```

使用已经在 RabbitMQ 中定义的 OpenStack 用户密码替换 RABBIT_PASS。

④ 在[DEFAULT]和[keystone_authtoken]项，配置 Keystone 身份认证服务组件访问。

```
[DEFAULT]
...
auth_strategy = keystone

[keystone_authtoken]
...
auth_uri = http://controller:5000
auth_url = http://controller:35357
auth_plugin = password
project_domain_id = default
user_domain_id = default
project_name = service
username = neutron
password = NEUTRON_PASS
```

使用已经定义的 Neutron 用户密码替换 NEUTRON_PASS。

> **注意**：
> 注释或者删除 [keystone_authtoken] 项中的其他内容。

⑤ 在[DEFAULT]和[nova]项，配置 Neutron 网络服务通知 Nova 计算服务关于网络拓扑的改变。

```
[DEFAULT]
...
notify_nova_on_port_status_changes = True
notify_nova_on_port_data_changes = True
nova_url = http://controller:8774/v2

[nova]
...
auth_url = http://controller:35357
auth_plugin = password
project_domain_id = default
user_domain_id = default
region_name = RegionOne
project_name = service
username = nova
password = NOVA_PASS
```

使用已经定义的 Nova 用户密码替换 NOVA_PASS。

⑥ 打开日志记录功能，方便问题跟踪和解决。

```
[DEFAULT]
...
verbose = True
```

（3）配置 Layer 2 (ML2) plug-in 模块。

ML2 plug-in 使用 bridge 机制为虚拟机实例桥接和交换网络信息。

① 编辑文件/etc/neutron/plugins/ml2/ml2_conf.ini，在[ml2]项，完成以下操作。

```
[ml2]
...
type_drivers = flat,vlan,vxlan
tenant_network_types = vxlan
mechanism_drivers = linuxbridge,l2population
```

> **错误**:
> 配置完成后,切勿删除 type_drivers 的值。

> **注意**:
> Neutron 网络服务组件中的桥接技术(bridge agent)仅支持 VXLAN 重叠网络(Overlay Network)技术。
>
> 目前主流重叠网络(Overlay Network)技术主要有 VXLAN、NVGRE 和 STT。
>
> **1. VXLAN**
>
> VXLAN 是将以太网报文封装在 UDP 传输层上的一种隧道转发模式,目的 UDP 端口号为 4798;为了使 VXLAN 充分利用承载网络路由的均衡性,VXLAN 通过将原始以太网数据头(MAC、IP、四层端口号等)的 Hash 值作为 UDP 的端口号;采用 24 比特标识二层网络分段,称为 VNI(VXLAN Network Identifier),类似于 VLAN ID 的作用;未知目的、广播、组播等网络流量均被封装为组播转发,物理网络要求支持任意源组播(ASM)。
>
> **2. NVGRE**
>
> NVGRE 是将以太网报文封装在 GRE 内的一种隧道转发模式;采用 24 比特标识二层网络分段,称为 VSI(Virtual Subnet Identifier),类似于 VLAN ID 的作用;为了使 NVGRE 利用承载网络路由的均衡性,NVGRE 在 GRE 扩展字段 flow ID,这就要求物理网络能够识别到 GRE 隧道的扩展信息,并以 flow ID 进行流量分担;未知目的、广播、组播等网络流量均被封装为组播转发。
>
> **3. STT**
>
> STT 利用了 TCP 的数据封装形式,但改造了 TCP 的传输机制,数据传输不遵循 TCP 状态机制,而是遵循全新定义的无状态机制,将 TCP 各字段意义重新定义,无须三次握手即可建立 TCP 连接,因此称为无状态 TCP;以太网数据封装在无状态 TCP;采用 64 比特 Context ID 标识二层网络分段;为了使 STT 充分利用承载网络路由的均衡性,通过将原始以太网数据头(MAC、IP、四层端口号等)的 Hash 值作为无状态 TCP 的源端口号;未知目的、广播、组播等网络流量均被封装为组播转发。

② 编辑文件/etc/neutron/plugins/ml2/ml2_conf.ini,在[ml2_type_flat]项,完成以下操作。

```
[ml2_type_flat]
```

```
...
flat_networks = public
```

③ 编辑文件/etc/neutron/plugins/ml2/ml2_conf.ini，在[ml2_type_vxlan]项，配置 VXLAN 为 Private Network 提供的标识范围。

```
[ml2_type_vxlan]
...
vni_ranges = 1:1000
```

④ 编辑文件/etc/neutron/plugins/ml2/ml2_conf.ini，在[securitygroup]项，启用 ipset 提高 Security Group 规则的效率。

```
[securitygroup]
...
enable_ipset = True
```

(4) 配置 bridge agent。

bridge agent 为虚拟机实例建立了 Layer 2（桥接和交换网络信息）虚拟网络，提供了实现 Private Network 所需的 VXLAN 和处理 Security Group。

① 编辑/etc/neutron/plugins/ml2/linuxbridge_agent.ini 文件，在[linux_bridge]项，将 Public Network 虚拟网络端口映射到物理网络端口。

```
[linux_bridge]
physical_interface_mappings = public:PUBLIC_INTERFACE_NAME
```

使用已经定义的物理网络端口替换 PUBLIC_INTERFACE_NAME 变量，针对此例，应该使用 eth1。

② 编辑文件 /etc/neutron/plugins/ml2/linuxbridge_agent.ini，在[vxlan]项，启用 VXLAN 功能和 Layer 2 population，配置处理 Overlay Network 的物理网络端口的 IP 地址。

```
[vxlan]
enable_vxlan = True
local_ip = OVERLAY_INTERFACE_IP_ADDRESS
l2_population = True
```

使用处理 Overlay Network 的物理端口 IP 地址替换 OVERLAY_INTERFACE_IP_ADDRESS 变量。针对本例，应该使用管理/数据网络的 IP 地址。

③ 编辑文件 /etc/neutron/plugins/ml2/linuxbridge_agent.ini，在[agent]项，启用 ARP 欺骗保护。

```
[agent]
...
prevent_arp_spoofing = True
```

④ 编辑/etc/neutron/plugins/ml2/linuxbridge_agent.ini 文件，在[securitygroup]项，启用 Security Group 和配置 iptables firewall driver。

```
[securitygroup]
...
enable_security_group = True
firewall_driver                                                     =
neutron.agent.linux.iptables_firewall.IptablesFirewallDriver
```

（5）配置 Layer 3 agent。

Layer 3 agent 为虚拟网络提供了路由（routing）和网络地址转换（NAT）服务。

① 编辑 /etc/neutron/l3_agent.ini 文件，在[DEFAULT]项，配置 bridge interface driver 和 external network bridge。

```
[DEFAULT]
...
interface_driver = neutron.agent.linux.interface.BridgeInterfaceDriver
external_network_bridge =
```

> 注意：
> "external_network_bridge"选项缺少值可以启用多个外部网络（External Network），而不必集中在某一个外部网络上。

② 在[DEFAULT]项，打开日志记录功能，方便问题跟踪和解决。

```
[DEFAULT]
...
verbose = True
```

（6）配置 DHCP agent。

DHCP agent 为虚拟网络提供了 DHCP 服务，也可以说为虚拟机实例提供了 IP 地址。

① 编辑/etc/neutron/dhcp_agent.ini 文件，在[DEFAULT]项，配置 bridge interface driver 和 Dnsmasq DHCP driver，启用 metadata 分离以使虚拟机实例访问互联网上的 metadata。

```
[DEFAULT]
...
interface_driver = neutron.agent.linux.interface.BridgeInterfaceDriver
dhcp_driver = neutron.agent.linux.dhcp.Dnsmasq
enable_isolated_metadata = True
```

② 打开日志记录功能，方便问题跟踪和解决。

```
[DEFAULT]
...
verbose = True
```

类似于 VXLAN 的 Overlay Network 技术利用软件定义技术修改了网络数据包头，增加了包头大小，减小了可承载用户数据空间的大小。虚拟机实例发送数据包使用 MTU（Maximum Transmission Unit）的默认大小（1500 Bytes）时，网络中的 PMTUD（Path MTU Discovery）技术会发现该类数据包，并且调整其大小进行传输。如果操作系统或者网络设备不支持 PMTUD 技术，将会导致网络环境的性能降低或者网络连接失败。

为了避免这一现象，可以在物理网络中启用 jumbo frame 功能，它支持最大 9000 Bytes 的 MTU 值，可以有效解决因 MTU 大小问题导致的性能降低和网络连接失败问题。尽管如此，由于一些网络设备不支持 jumbo frame 功能或者 OpenStack 项目的管理员缺少相关的网络知识，可能还是会存在问题。在此，有另外一种解决办法，即调整 MTU 值，在大多数情况下，MTU 值调整为 1450 Bytes 是有效的。

注意：
　　一些云中的镜像忽略了调整 MTU 值大小的问题，也可以通过 metadata、脚本或其他办法进行补救。

③ 编辑 /etc/neutron/dhcp_agent.ini 文件，在[DEFAULT]项，定义 Dnsmasq 配置文件的路径。

```
[DEFAULT]
...
dnsmasq_config_file = /etc/neutron/dnsmasq-neutron.conf
```

④ 创建并编辑/etc/neutron/dnsmasq-neutron.conf 文件，设置 MTU 值的大小。

```
dhcp-option-force=26,1450
```

### 14.1.3 配置 metadata agent

metadata agent 对虚拟机实例提供一些配置信息，如凭据（credential）。

编辑 /etc/neutron/metadata_agent.ini 文件，完成以下操作。

（1）在[DEFAULT]项，配置 Neutron 用户的访问。

```
[DEFAULT]
...
auth_uri = http://controller:5000
auth_url = http://controller:35357
auth_region = RegionOne
auth_plugin = password
project_domain_id = default
user_domain_id = default
project_name = service
username = neutron
password = NEUTRON_PASS
```

使用已经定义的 Neutron 用户密码替换 NEUTRON_PASS。

（2）在[DEFAULT]项，配置 metadata 相关信息。

```
[DEFAULT]
...
nova_metadata_ip = Controller
metadata_proxy_shared_secret = METADATA_SECRET
```

使用合适的一串字段代替 METADATA_SECRET 变量，当然也可以默认使用 METADATA_SECRET 字段。

(3) 打开日志记录功能，方便问题跟踪和解决。

```
[DEFAULT]
...
verbose = True
```

## 14.1.4 配置计算服务组件

编辑 /etc/nova/nova.conf 文件，完成以下步骤。

在[neutron]项，配置 Neutron 访问参数，配置 metadata 相关信息。

```
[neutron]
...
url = http://controller:9696
auth_url = http://controller:35357
auth_plugin = password
project_domain_id = default
user_domain_id = default
region_name = RegionOne
project_name = service
username = neutron
password = NEUTRON_PASS

service_metadata_proxy = True
metadata_proxy_shared_secret = METADATA_SECRET
```

使用已经定义的 Neutron 用户密码替换 NEUTRON_PASS。

使用已经定义的 metadata proxy 密码代替 METADATA_SECRET。

## 14.1.5 安装完成

（1）Neutron 网络服务组件初始化脚本中的参数 NEUTRON_PLUGIN_CONF 需要关联 ML2 plug-in 配置文件。编辑/etc/sysconfig/neutron 文件，添加以下内容。

```
NEUTRON_PLUGIN_CONF="/etc/neutron/plugins/ml2/ml2_conf.ini"
```

(2) 重启 Nova 计算服务的 API 服务。

```
# systemctl restart openstack-nova-api.service
```

(3) 启动 Neutron 网络服务组件并设置为开机自启动（无论采用架构一或架构二，都需要启动以下服务）。

```
# systemctl enable openstack-neutron.service \
  openstack-neutron-linuxbridge-agent.service \
  openstack-neutron-dhcp-agent.service \
  openstack-neutron-metadata-agent.service
# systemctl start openstack-neutron.service \
  openstack-neutron-linuxbridge-agent.service \
  openstack-neutron-dhcp-agent.service \
  openstack-neutron-metadata-agent.service
```

(4) 如果采用架构二，则启动以下服务并设置为开机自启动。

```
# systemctl enable openstack-neutron-l3-agent.service
# systemctl start openstack-neutron-l3-agent.service
```

## 14.2 安装和配置（计算节点）

本节详细描述如何在计算节点上安装和配置 Neutron 网络服务组件，计算节点为虚拟机实例提供网络连接和安全组（Security Group）功能。

### 14.2.1 网络服务组件安装和配置通用组件

下面介绍安装 Neutron 网络服务组件的计算节点部分，同时对关于 Neutron 网络服务组件的通用部分进行配置。

(1) 安装软件包。

```
# zypper install --no-recommends openstack-neutron-linuxbridge-agent ipset
```

(2) 配置通用组件。

Neutron 网络服务组件的通用部分包括认证机制、消息队列和 plug-in。

① 编辑/etc/neutron/neutron.conf 文件，在[database]项，注释掉任何数据库访问连接配置。不允许计算节点 s 直接访问数据库，必须通过 nova-conductor module 间接访问。

② 编辑/etc/neutron/neutron.conf 文件，在[DEFAULT]和[oslo_messaging_rabbit]项，配置 RabbitMQ 消息队列的访问。

```
[DEFAULT]
...
rpc_backend = rabbit

[oslo_messaging_rabbit]
...
rabbit_host = controller
rabbit_userid = openstack
rabbit_password = RABBIT_PASS
```

使用已经定义的 OpenStack 用户密码替换 RABBIT_PASS。

③ 编辑/etc/neutron/neutron.conf 文件，在[DEFAULT]和[keystone_authtoken]项，配置 Keystone 身份认证服务组件访问。

```
[DEFAULT]
...
auth_strategy = keystone

[keystone_authtoken]
...
auth_uri = http://controller:5000
auth_url = http://controller:35357
auth_plugin = password
project_domain_id = default
user_domain_id = default
project_name = service
username = neutron
password = NEUTRON_PASS
```

使用已经定义的 Neutron 用户密码替换 NEUTRON_PASS。

**注意**：

注释或者删除[keystone_authtoken]项中的其他内容。

④ 打开日志记录功能，方便问题跟踪和解决。

```
[DEFAULT]
...
verbose = True
```

## 14.2.2　配置网络核心组件

选择与控制节点上相同架构的网络配置进行操作。

### 1．架构一配置（Provider Network）

以下操作均在控制节点上完成。

bridge agent 为虚拟机实例建立了 Layer 2（桥接和交换网络信息）虚拟网络，提供了实现 Private Network 所需的 VXLAN 和处理 Security Group。

（1）编辑/etc/neutron/plugins/ml2/linuxbridge_agent.ini 文件，在[linux_bridge]项，将 Public Network 虚拟网络端口映射到物理网络端口。

```
[linux_bridge]
physical_interface_mappings = public:PUBLIC_INTERFACE_NAME
```

使用已经定义的物理网络端口替换 PUBLIC_INTERFACE_NAME，针对此例，应该是 eth1。

（2）编辑/etc/neutron/plugins/ml2/linuxbridge_agent.ini 文件，在[vxlan]项，禁用 VXLAN Overlay Network。

```
[vxlan]
enable_vxlan = False
```

（3）编辑/etc/neutron/plugins/ml2/linuxbridge_agent.ini 文件，在[agent]项，启用 ARP 欺骗保护。

```
[agent]
...
prevent_arp_spoofing = True
```

（4）编辑/etc/neutron/plugins/ml2/linuxbridge_agent.ini 文件，在[securitygroup]项，启用 Security Group 和配置 iptables firewall driver。

```
[securitygroup]
...
enable_security_group = True
firewall_driver                                                             =
neutron.agent.linux.iptables_firewall.IptablesFirewallDriver
```

### 2．架构二配置（Self-Service Network）

以下操作均在控制节点上完成。

bridge agent 为虚拟机实例建立了 Layer 2（桥接和交换网络信息）虚拟网络，提供了实现 Private Network 所需的 VXLAN 和处理 Security Group。

（1）编辑/etc/neutron/plugins/ml2/linuxbridge_agent.ini 文件，在[linux_bridge]项，将 Public Network 虚拟网络端口映射到物理网络端口。

```
[linux_bridge]
physical_interface_mappings = public:PUBLIC_INTERFACE_NAME
```

使用已经定义的物理网络端口替换 PUBLIC_INTERFACE_NAME，针对此例，应该是 eth1。

（2）编辑/etc/neutron/plugins/ml2/linuxbridge_agent.ini 文件，在[vxlan]项，启用 VXLAN 功能和 Layer 2 population，配置处理 Overlay Network 的物理网络端口的 IP 地址。

```
[vxlan]
enable_vxlan = True
local_ip = OVERLAY_INTERFACE_IP_ADDRESS
l2_population = True
```

使用处理 Overlay Network 的物理端口 IP 地址替换 OVERLAY_INTERFACE_IP_ADDRESS 变量。针对本例，应该使用管理/数据网络的 IP 地址。

(3) 编辑/etc/neutron/plugins/ml2/linuxbridge_agent.ini 文件，在[agent]项，启用 ARP 欺骗保护。

```
[agent]
...
prevent_arp_spoofing = True
```

(4) 编辑/etc/neutron/plugins/ml2/linuxbridge_agent.ini 文件，在[securitygroup]项，启用 Security Group 和配置 iptables firewall driver。

```
[securitygroup]
...
enable_security_group = True
firewall_driver                                                          = neutron.agent.linux.iptables_firewall.IptablesFirewallDriver
```

## 14.2.3 配置计算服务组件

编辑 /etc/nova/nova.conf 文件，完成以下步骤。

在[neutron]项，配置 Neutron 访问参数。

```
[neutron]
...
url = http://controller:9696
auth_url = http://controller:35357
auth_plugin = password
project_domain_id = default
user_domain_id = default
region_name = RegionOne
project_name = service
username = neutron
password = NEUTRON_PASS
```

使用已经定义的 Neutron 用户密码替换 NEUTRON_PASS。

## 14.2.4 安装完成

（1）Neutron 网络服务组件初始化脚本中的参数 NEUTRON_PLUGIN_CONF 需要关联 ML2 plug-in 配置文件。编辑/etc/sysconfig/neutron 文件，添加以下内容。

```
NEUTRON_PLUGIN_CONF="/etc/neutron/plugins/ml2/ml2_conf.ini"
```

（2）重启 Nova 计算服务组件。

```
# systemctl restart openstack-nova-compute.service
```

（3）启动 bridge agent 并设置为开机自启动。

```
# systemctl enable openstack-neutron-linuxbridge-agent.service
# systemctl start openstack-neutron-linuxbridge-agent.service
```

## 14.3 验证

下面检查一下配置完成的 Neutron 网络服务组件。所有的操作均在控制节点上进行。

（1）加载 admin-openrc.sh 文件，自动生成环境变量。

```
# source admin-openrc.sh
```

（2）列出所有选项，以确认是否成功启动 neutron-server 进程。

```
# neutron ext-list
+-----------------------+---------------------------+
| alias                 | name                      |
+-----------------------+---------------------------+
| flavors               | Neutron Service Flavors   |
| security-group        | security-group            |
| dns-integration       | DNS Integration           |
| net-mtu               | Network MTU               |
| port-security         | Port Security             |
| binding               | Port Binding              |
| provider              | Provider Network          |
| agent                 | agent                     |
| quotas                | Quota management support  |
| subnet_allocation     | Subnet Allocation         |
| dhcp_agent_scheduler  | DHCP Agent Scheduler      |
| rbac-policies         | RBAC Policies             |
| external-net          | Neutron external network  |
| multi-provider        | Multi Provider Network    |
| allowed-address-pairs | Allowed Address Pairs     |
| extra_dhcp_opt        | Neutron Extra DHCP opts   |
+-----------------------+---------------------------+
```

（3）采用架构一进行配置，查看 Neutron agents 是否成功启动。

```
# neutron agent-list
+--------------------------------------+--------------------+------------+-------+----------------+
| id                                   | agent_type         | host       | alive | admin_state_up |
+--------------------------------------+--------------------+------------+-------+----------------+
| 41648844-fba3-4bf7-afaa-e98db4c7f8bb | Linux bridge agent | Controller | :-)   | True           |
| 416d377b-d1ff-4637-b16b-2bd5bcab8896 | DHCP agent         | Controller | :-)   | True           |
| 5a8a0625-2a6e-4fa9-a015-79ade84a9ca4 | Metadata agent     | Controller | :-)   | True           |
| d632bc03-fced-4ed0-a078-bb90594d0f93 | Linux bridge agent | Compute02  | :-)   | True           |
| ec7fb81e-4475-45af-a81d-16d999a7a025 | Linux bridge agent | Compute01  | :-)   | True           |
+--------------------------------------+--------------------+------------+-------+----------------+
```

采用架构二进行配置，查看 Neutron agents 是否成功启动（架构二比架构一多一个 L3 agent）。

```
# neutron agent-list
+--------------------------------------+--------------------+------------+-------+----------------+
| id                                   | agent_type         | host       | alive | admin_state_up |
+--------------------------------------+--------------------+------------+-------+----------------+
| 08905043-5010-4b87-bba5-aedb1956e27a | Linux bridge agent | compute1   | :-)   | True           |
| 27eee952-a748-467b-bf71-941e89846a92 | Linux bridge agent | controller | :-)   | True           |
| 830344ff-dc36-4956-84f4-067af667a0dc | L3 agent           | controller | :-)   | True           |
| dd3644c9-1a3a-435a-9282-eb306b4b0391 | DHCP agent         | controller | :-)   | True           |
| f49a4b81-afd6-4b3d-b923-66c8f0517099 | Metadata agent     | controller | :-)   | True           |
+--------------------------------------+--------------------+------------+-------+----------------+
```

# 第 15 章 仪表板（Horizon）服务安装配置

OpenStack 项目中的 Horizon 仪表板服务组件是以 Web 界面形式展示各项服务的，OpenStack 云系统管理员和终端用户可以通过仪表板管理各项资源和服务。

Horizon 仪表板服务组件通过 OpenStack 的 APIs 与控制节点通信，分配资源和同步状态。

Horizon 仪表板服务组件允许对其进行定制化的修改，同时，提供了一些核心代码类和可重复使用的模板与工具。

本章的 Horizon 仪表板服务组件部署使用 Apache Web Server。

## 15.1 安装和配置

本节详细描述如何在控制节点上安装 Horizon 仪表板服务组件。如之前所述，Horizon 仪表板服务组件是基于 Web 界面形式展现的，多个核心功能服务组件依附于它进行展示，如 Keystone 身份认证服务组件、Glance 镜像服务组件、Nova 计算服务组件、Neutron 网络服务组件或者已经不流行的 nova-network 服务组件等。当然，一些服务组件也不能使用 Horizon 仪表板服务组件，如 Swift 对象存储。

**注意**：

本章安装和配置 Horizon 仪表板服务组件的正确性是建立在之前 Keystone 身份认证服务组件正确安装和配置的基础上的，如果 Keystone 身份认证服务组件没有被正确安装与配置，则 Horizon 仪表板服务组件也会存在问题。

## 15.1.1 安装和配置 Horizon 仪表板服务组件

下面开始安装和配置 Horizon 仪表板服务组件，由于版本的不同，可能一些配置文件的内容也不同，需要修改或者添加一些内容。省略号（...）代表配置文件中需要保留的默认配置信息。

（1）安装软件包。

```
# zypper install openstack-dashboard
```

（2）配置 Web 应用服务器。

```
# cp /etc/apache2/conf.d/openstack-dashboard.conf.sample \
  /etc/apache2/conf.d/openstack-dashboard.conf
# a2enmod rewrite;a2enmod ssl;a2enmod wsgi
```

（3）编辑/srv/www/openstack-dashboard/openstack_dashboard/local/local_settings.py 文件，完成以下操作。

① 配置 Horizon 仪表板服务组件在控制节点上使用其他服务组件。

```
OPENSTACK_HOST = "Controller"
```

② 允许所有 IP 访问 Horizon 仪表板服务组件。

```
ALLOWED_HOSTS = ['*', ]
```

③ 配置 MemCached。

```
CACHES = {
    'default': {
        'BACKEND':
'django.core.cache.backends.memcached.MemcachedCache',
        'LOCATION': '127.0.0.1:11211',
```

    }
}

 **注意：**

注释其他存储配置。

④ 配置 user 作为通过 dashboard 创建用户的默认权限。

```
OPENSTACK_KEYSTONE_DEFAULT_ROLE = "user"
```

⑤ 支持使用多个 Domain。

```
OPENSTACK_KEYSTONE_MULTIDOMAIN_SUPPORT = True
```

⑥ 配置服务组件的 API 版本。

```
OPENSTACK_API_VERSIONS = {
    "identity": 3,
    "volume": 2,
}
```

⑦ 如果在 Neutron 网络服务组件部分选择的是架构一建设方案，则需要禁用 Layer 3 服务。

```
OPENSTACK_NEUTRON_NETWORK = {
    ...
    'enable_router': False,
    'enable_quotas': False,
    'enable_distributed_router': False,
    'enable_ha_router': False,
    'enable_lb': False,
    'enable_firewall': False,
    'enable_vpn': False,
    'enable_fip_topology_check': False,
}
```

⑧ 配置时区。

```
TIME_ZONE = "TIME_ZONE"
```

使用合适的时区替换 TIME_ZONE 字段。

## 15.1.2 安装完成

启动服务并设置为开机自启动。

```
# systemctl enable apache2.service memcached.service
# systemctl restart apache2.service memcached.service
```

## 15.2 验证

下面检查一下配置完成的 Horizon 仪表板服务组件。打开浏览器，输入 http://Controller_ip/，可以使用 admin 或者 demo 用户登录，密码与用户名相同。

注意：
　　至此，我们可以使用建立虚拟机实例、分配 IP 地址、定制化 Horizon 仪表板、使用 VNC 进行本地登录等功能，一个初步的 OpenStack 云环境模型已经呈现。

# 第 16 章
# 块存储（Cinder）服务安装配置

OpenStack 项目中的 Cinder 块存储服务组件通过运行在主机节点上的多个 cinder-* 进程对外提供服务，其可以运行在一个或多个存储节点上，也可以与其他服务组件共同运行在同一主机节点上。

为了更好地理解和管理 Cinder 块存储服务组件，需要对其整个存储架构有一定的了解，在经济、性能和安全性方面做出选择，但是最终归结为两个选择：是单节点部署还是多节点部署（在存储技术已经确定的情况下）。Cinder 块存储服务组件允许用户添加外部存储作为块存储使用，如集中存储或其他开源存储技术存储。Amazon 的 EC2 Elastic 块存储（EBS）与其类似。

## 16.1 安装和配置（控制节点）

本节详细描述如何在控制节点上安装和配置 Cinder 块存储服务组件。Cinder 块存储服务组件要求至少有一个存储节点。

## 16.1.1 准备

在安装和配置 Cinder 块存储服务组件之前,首先创建数据库、服务证书(service credential)和 API endpoint。

(1) 创建数据库。

① 在操作系统终端连接数据库。

```
# mysql -u root -p
```

因为我们使用 root 用户登录,所以在提示输入密码时,按回车键即可建立连接。

② 创建 Cinder 数据库。

```
MariaDB [(none)]> CREATE DATABASE cinder;
```

③ Cinder 数据库的访问权限设置。

```
GRANT ALL PRIVILEGES ON cinder.* TO 'cinder'@'localhost' \
  IDENTIFIED BY 'CINDER_DBPASS';
GRANT ALL PRIVILEGES ON cinder.* TO 'cinder'@'%' \
  IDENTIFIED BY 'CINDER_DBPASS';
```

替换 CINDER_DBPASS 为合适的密码。

④ 退出数据库。

(2) 加载 admin user 的环境变量。

```
# source admin-openrc.sh
```

(3) 创建 Identity 服务凭据。

① 创建 Cinder 用户。

```
# openstack user create --domain default --password-prompt cinder
User Password:
Repeat User Password:
```

```
+-----------+------------------------------------+
| Field     | Value                              |
+-----------+------------------------------------+
| domain_id | default                            |
| enabled   | True                               |
| id        | 0c054bcf0bf54016966c5f096cc05add   |
| name      | cinder                             |
+-----------+------------------------------------+
```

② 将 admin role 赋予 Cinder 用户和 service project。

```
# openstack role add --project service --user cinder admin
```

 注意：

此命令执行完成后，没有输出。

③ 创建 Cinder 和 Cinder v2 service entity。

```
# openstack service create --name cinder \
  --description "OpenStack Block storage" volume
+-------------+----------------------------------+
| Field       | Value                            |
+-------------+----------------------------------+
| description | OpenStack Block Storage          |
| enabled     | True                             |
| id          | 7afa75c21c6e4405a7766b9bba89c985 |
| name        | cinder                           |
| type        | volume                           |
+-------------+----------------------------------+

# openstack service create --name cinderv2 \
  --description "OpenStack Block storage" volumev2
+-------------+----------------------------------+
| Field       | Value                            |
+-------------+----------------------------------+
| description | OpenStack Block Storage          |
| enabled     | True                             |
| id          | 85b43066b586458fa06b39c0d80fd10a |
| name        | cinderv2                         |
| type        | volumev2                         |
+-------------+----------------------------------+
```

 注意：

块存储要求有两种 service entity。

（4）创建 Cinder 块存储服务组件的 API endpoint。

```
# openstack endpoint create --region RegionOne \
  volume public http://controller:8776/v1/%\(tenant_id\)s
```

```
+--------------+------------------------------------------+
| Field        | Value                                    |
+--------------+------------------------------------------+
| enabled      | True                                     |
| id           | 0be30557f91d4de5bd5e21ea46118d2c         |
| interface    | public                                   |
| region       | RegionOne                                |
| region_id    | RegionOne                                |
| service_id   | 7afa75c21c6e4405a7766b9bba89c985         |
| service_name | cinder                                   |
| service_type | volume                                   |
| url          | http://Controller:8776/v1/%(tenant_id)s  |
+--------------+------------------------------------------+

# openstack endpoint create --region RegionOne \
  volume internal http://controller:8776/v1/%\(tenant_id\)s

+--------------+------------------------------------------+
| Field        | Value                                    |
+--------------+------------------------------------------+
| enabled      | True                                     |
| id           | 047acefe3006468b92cb4309f9642800         |
| interface    | internal                                 |
| region       | RegionOne                                |
| region_id    | RegionOne                                |
| service_id   | 7afa75c21c6e4405a7766b9bba89c985         |
| service_name | cinder                                   |
| service_type | volume                                   |
| url          | http://Controller:8776/v1/%(tenant_id)s  |
+--------------+------------------------------------------+

# openstack endpoint create --region RegionOne \
  volume admin http://controller:8776/v1/%\(tenant_id\)s

+--------------+------------------------------------------+
| Field        | Value                                    |
+--------------+------------------------------------------+
| enabled      | True                                     |
| id           | 3803bb39fb554ceebda8a005a579e204         |
| interface    | admin                                    |
| region       | RegionOne                                |
| region_id    | RegionOne                                |
| service_id   | 7afa75c21c6e4405a7766b9bba89c985         |
| service_name | cinder                                   |
| service_type | volume                                   |
| url          | http://Controller:8776/v1/%(tenant_id)s  |
+--------------+------------------------------------------+

# openstack endpoint create --region RegionOne \
  volumev2 public http://controller:8776/v2/%\(tenant_id\)s

+--------------+------------------------------------------+
| Field        | Value                                    |
+--------------+------------------------------------------+
| enabled      | True                                     |
| id           | 19cab8254a8d42769fe54fc9b80a9fc3         |
| interface    | public                                   |
| region       | RegionOne                                |
| region_id    | RegionOne                                |
| service_id   | 85b43066b586458fa06b39c0d80fd10a         |
| service_name | cinderv2                                 |
| service_type | volumev2                                 |
| url          | http://Controller:8776/v2/%(tenant_id)s  |
+--------------+------------------------------------------+
```

```
# openstack endpoint create --region RegionOne \
  volumev2 internal http://controller:8776/v2/%\(tenant_id\)s
+--------------+------------------------------------------+
| Field        | value                                    |
+--------------+------------------------------------------+
| enabled      | True                                     |
| id           | 0fb58a962b0d481aabe2faa9685f6c1c         |
| interface    | internal                                 |
| region       | RegionOne                                |
| region_id    | RegionOne                                |
| service_id   | 85b43066b586458fa06b39c0d80fd10a         |
| service_name | cinderv2                                 |
| service_type | volumev2                                 |
| url          | http://Controller:8776/v2/%(tenant_id)s  |
+--------------+------------------------------------------+

# openstack endpoint create --region RegionOne \
  volumev2 admin http://controller:8776/v2/%\(tenant_id\)s
+--------------+------------------------------------------+
| Field        | value                                    |
+--------------+------------------------------------------+
| enabled      | True                                     |
| id           | 1e5c9316f06b462e84f79c80c5250d01         |
| interface    | admin                                    |
| region       | RegionOne                                |
| region_id    | RegionOne                                |
| service_id   | 85b43066b586458fa06b39c0d80fd10a         |
| service_name | cinderv2                                 |
| service_type | volumev2                                 |
| url          | http://Controller:8776/v2/%(tenant_id)s  |
+--------------+------------------------------------------+
```

注意：

块存储要求为两个 service entity 分别创建 endpoint。

## 16.1.2 安装和配置 Cinder 块存储服务组件

下面开始安装和配置 Cinder 块存储服务组件，由于版本的不同，可能一些配置文件的内容也不同，需要修改或者添加一些内容。省略号（…）代表配置文件中需要保留的默认配置信息。

（1）安装软件包。

```
# zypper install openstack-cinder-api openstack-cinder-scheduler python-cinderclient
```

（2）编辑/etc/cinder/cinder.conf 文件，完成以下操作。

① 在[database]项，配置数据库连接。

```
[database]
...
connection = mysql+pymysql://cinder:CINDER_DBPASS@controller/cinder
```

使用已经定义的 Cinder 数据库密码替换 CINDER_DBPASS。

② 在[DEFAULT]和[oslo_messaging_rabbit]项，配置 RabbitMQ 消息队列连接。

```
[DEFAULT]
...
rpc_backend = rabbit

[oslo_messaging_rabbit]
...
rabbit_host = controller
rabbit_userid = openstack
rabbit_password = RABBIT_PASS
```

使用已经在 RabbitMQ 中定义的 OpenStack 用户密码替换 RABBIT_PASS。

③ 在[DEFAULT]和[keystone_authtoken]项，配置 Keystone 身份认证服务组件访问。

```
[DEFAULT]
...
auth_strategy = keystone

[keystone_authtoken]
...
auth_uri = http://controller:5000
auth_url = http://controller:35357
auth_plugin = password
project_domain_id = default
user_domain_id = default
project_name = service
username = cinder
password = CINDER_PASS
```

使用已经定义的 Cinder 用户密码替换 CINDER_PASS。

**注意:**

注释或者删除[keystone_authtoken]项中的其他内容。

④ 在[DEFAULT]项,配置 my_ip 参数为控制节点的管理/数据网络 IP 地址。

```
[DEFAULT]
...
my_ip = 172.168.1.11
```

⑤ 在[oslo_concurrency]项,配置锁路径。

```
[oslo_concurrency]
...
lock_path = /var/lib/cinder/tmp
```

⑥ 打开日志记录功能,方便问题跟踪和解决。

```
[DEFAULT]
...
verbose = True
```

(3) 将 Cinder 块存储服务信息同步到 Cinder 数据库中。

```
# su -s /bin/sh -c "cinder-manage db sync" cinder
```

(4) 编辑/etc/nova/nova.conf 文件,配置 Nova 计算服务使用 Cinder 块存储服务。

```
[cinder]
os_region_name = RegionOne
```

## 16.1.3 安装完成

(1) 重启 Nova 计算服务的 API 服务。

```
# systemctl restart openstack-nova-api.service
```

(2) 启动 Cinder 块存储服务组件并设置为开机自启动。

```
# systemctl enable openstack-cinder-api.service openstack-cinder-scheduler.service
# systemctl start openstack-cinder-api.service openstack-cinder-scheduler.service
```

## 16.2 安装和配置（存储节点）

本节详细描述如何在存储节点上安装和配置 Cinder 块存储服务组件。简单来说，本次配置将计算节点和存储节点合二为一，放在同一主机节点上，使用本地硬件设备（/dev/sdb）作为存储，通过 iSCSI 传输协议，使用 LVM 驱动将逻辑卷分配给虚拟机实例。

### 16.2.1 准备

在安装和配置 Cinder 块存储服务组件之前，需要首先确定使用的存储设备。以下操作均在存储节点上进行。

（1）安装软件包。

① 安装 lvm2 软件包。

```
# zypper install lvm2
```

 注意：

在安装 lvm2 之前，可以通过以下命令先检查是否已经安装：
```
rpm -qa|grep lvm2
```
lvm2 是操作系统的标准组件，一般默认安装。

② 安装 qemu 软件包。使用 non-raw 镜像类型（如 vmdk、QCOW2）时需要使用。

```
# zypper install qemu
```

（2）创建物理卷/dev/sdb。

```
# pvcreate /dev/sdb
Physical volume "/dev/sdb" successfully created
```

（3）创建卷组 cinder-volumes。

```
# vgcreate cinder-volumes /dev/sdb
Volume group "cinder-volumes" successfully created
```

（4）Cinder 块存储卷一般只能被虚拟机实例访问使用，但是存储节点操作系统可以管理包括磁盘在内的本地硬件设备，操作系统中的 LVM 卷扫描工具可以扫描/dev 目录下的所有设备,包括虚拟机实例访问使用的块存储卷。如果块存储卷被虚拟机实例用作 LVM,则当卷扫描工具尝试对块存储卷进行扫描时,会产生一系列问题。为了避免这些问题,需要对块存储卷进行重配置。编辑/etc/lvm/lvm.conf 文件,完成以下操作。

在 devices 项,添加一个过滤条目。

```
devices {
...
filter = [ "a/sdb/", "r/.*/"]
```

每个过滤条目(filter)都以 a(accept)或 r(reject)开始,包含一个硬件设备,必须以"r/.*/"结束,vgs –vvvv 可以测试过滤条目是否生效。

 警告:

如果存储节点的操作系统所在磁盘使用 LVM,则必须把该磁盘加入到过滤条目中。例如,在/dev/sda 磁盘上安装了操作系统:

```
filter = [ "a/sda/", "a/sdb/", "r/.*/"]
```

同样,如果计算节点的操作系统所在磁盘使用 LVM,则也必须将磁盘加入到过滤条目中。例如,在/dev/sda 磁盘上安装了操作系统:

```
filter = [ "a/sda/", "r/.*/"]
```

## 16.2.2 安装和配置 Cinder 块存储服务组件

(1)安装软件包。

```
# zypper install openstack-cinder-volume tgt python-PyMySQL
```

(2)编辑/etc/cinder/cinder.conf 文件,完成以下操作。

① 在[database]项,配置数据库连接。

```
[database]
...
connection = mysql+pymysql://cinder:CINDER_DBPASS@controller/cinder
```

使用已经定义的 Cinder 数据库密码替换 CINDER_DBPASS。

② 在[DEFAULT]和[oslo_messaging_rabbit]项，配置 RabbitMQ 消息队列连接。

```
[DEFAULT]
...
rpc_backend = rabbit

[oslo_messaging_rabbit]
...
rabbit_host = controller
rabbit_userid = openstack
rabbit_password = RABBIT_PASS
```

使用已经在 RabbitMQ 中定义的 OpenStack 用户密码替换 RABBIT_PASS。

③ 在[DEFAULT]和[keystone_authtoken]项，配置 Keystone 身份认证服务组件访问。

```
[DEFAULT]
...
auth_strategy = keystone

[keystone_authtoken]
...
auth_uri = http://controller:5000
auth_url = http://controller:35357
auth_plugin = password
project_domain_id = default
user_domain_id = default
project_name = service
username = cinder
password = CINDER_PASS
```

使用已经定义的 Cinder 用户密码替换 CINDER_PASS。

注意：

注释或者删除[keystone_authtoken]项中的其他内容。

④ 在[DEFAULT]项，配置 my_ip 参数为控制节点的管理/数据网络 IP 地址。

```
[DEFAULT]
...
```

```
my_ip = 172.168.1.12
```

使用存储节点上管理/数据网络的 IP 地址替换 my_ip 字段。

⑤ 在[lvm]项，配置驱动类型、卷组名称和通信协议。

```
[lvm]
...
volume_driver = cinder.volume.drivers.lvm.LVMVolumeDriver
volume_group = cinder-volumes
iscsi_protocol = iscsi
iscsi_helper = tgtadm
```

⑥ 在[DEFAULT]项，设置后端名称。

```
[DEFAULT]
...
enabled_backends = lvm
```

**注意：**
后端名称是随意定义的。

⑦ 在[DEFAULT]项，配置 Image 服务组件的位置。

```
[DEFAULT]
...
glance_host = Controller
```

⑧ 在[oslo_concurrency]项，配置锁路径。

```
[oslo_concurrency]
...
lock_path = /var/lib/cinder/tmp
```

⑨ 打开日志记录功能，方便问题跟踪和解决。

```
[DEFAULT]
...
verbose = True
```

## 16.2.3 安装完成

启动块存储卷的相关服务并设置为开机自启动。

```
# systemctl enable openstack-cinder-volume.service tgtd.service
# systemctl start openstack-cinder-volume.service tgtd.service
```

## 16.3 验证

下面检查一下配置完成的 Cinder 块存储服务组件。所有的操作均在控制节点上进行。

（1）加载 admin-openrc.sh 文件，自动生成环境变量。

```
# source admin-openrc.sh
```

（2）列出 Cinder 块存储服务组件，看进程是否成功启动。

```
# cinder service-list
+------------------+----------------+------+---------+-------+----------------------------+
|      Binary      |      Host      | Zone | Status  | State |         Updated_at         |
+------------------+----------------+------+---------+-------+----------------------------+
| cinder-scheduler |   Controller   | nova | enabled |   up  | 2016-04-20T07:03:05.000000 |
|  cinder-volume   | Compute01@lvm  | nova | enabled |   up  | 2016-04-20T07:03:06.000000 |
|  cinder-volume   | Compute02@lvm  | nova | enabled |   up  | 2016-04-20T07:04:04.000000 |
+------------------+----------------+------+---------+-------+----------------------------+
```

# 第 17 章 对象存储（Swift）服务安装配置

Swift 对象存储服务组件利用标准的 x86 架构服务器组成集群，提供具备冗余性和可扩展性的数据存储。它属于持久性存储，可以长期保存数据，支持对数据的检索和更新功能。Swift 对象存储服务组件使用分布式架构，没有控制单点，提供更高的冗余性、扩展性和性能。数据被保存到多个主机节点的多个硬件设备上，由软件负责进行数据复制和数据容错，当主机节点不可用时，该主机节点上的数据会自动复制到其他可用的主机节点上。Swift 对象存储服务组件支持横向扩展，提供完全分布式的架构和 API 调用即可访问的存储平台，可以与各种应用软件集成，用于数据备份、归档和保存。

## 17.1 安装和配置（控制节点）

本书详细描述如何在控制节点上安装和配置 Proxy 服务，它可以处理 Account、Container 和 Object 在存储节点操作的请求。当然，Proxy 服务可以安装在任意能够连接存储节点的服务器上，为了提高性能和冗余性，可以在多个服务器上安装和配置 Proxy 服务。

## 17.1.1 准备

在 OpenStack 项目中，Proxy 服务依赖于 Keystone 身份认证服务组件，需要 Keystone 身份认证服务组件提供的认证和授权机制。同时，它区别于其他服务组件，Swift 对象存储拥有自己的内部存储机制，独立使用，不依赖于其他服务组件。

>  **注意**：
> Swift 对象存储不使用控制节点的 SQL 数据库，而是使用每个存储节点上的分布式 SQLite 数据库。

（1）加载 admin user 的环境变量。

```
# source admin-openrc.sh
```

（2）创建 Identity 服务凭据。

① 创建 Swift 用户。

```
# openstack user create --domain default --password-prompt swift
User Password:
Repeat User Password:
```

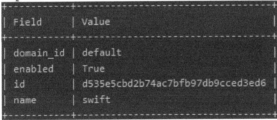

② 将 admin role 赋予 swift user 和 service project。

```
# openstack role add --project service --user swift admin
```

>  **注意**：
> 此命令执行完成后，没有输出。

③ 创建 swift service entity。

```
# openstack service create --name swift \
  --description "OpenStack Object storage" object-store
```

```
+-------------+------------------------------------+
| Field       | Value                              |
+-------------+------------------------------------+
| description | OpenStack Object Storage           |
| enabled     | True                               |
| id          | 75ef509da2c340499d454ae96a2c5c34   |
| name        | swift                              |
| type        | object-store                       |
+-------------+------------------------------------+
```

(3) 创建 Swift 对象存储服务组件的 API endpoint。

```
# openstack endpoint create --region RegionOne \
  object-store public http://controller:8080/v1/AUTH_%\(tenant_id\)s
+--------------+----------------------------------------------+
| Field        | Value                                        |
+--------------+----------------------------------------------+
| enabled      | True                                         |
| id           | 12bfd36f26694c97813f665707114e0d             |
| interface    | public                                       |
| region       | RegionOne                                    |
| region_id    | RegionOne                                    |
| service_id   | 75ef509da2c340499d454ae96a2c5c34             |
| service_name | swift                                        |
| service_type | object-store                                 |
| url          | http://controller:8080/v1/AUTH_%(tenant_id)s |
+--------------+----------------------------------------------+

# openstack endpoint create --region RegionOne \
  object-store internal http://controller:8080/v1/AUTH_%\(tenant_id\)s
+--------------+----------------------------------------------+
| Field        | Value                                        |
+--------------+----------------------------------------------+
| enabled      | True                                         |
| id           | 7a36bee6733a4b5590d74d3080ee6789             |
| interface    | internal                                     |
| region       | RegionOne                                    |
| region_id    | RegionOne                                    |
| service_id   | 75ef509da2c340499d454ae96a2c5c34             |
| service_name | swift                                        |
| service_type | object-store                                 |
| url          | http://controller:8080/v1/AUTH_%(tenant_id)s |
+--------------+----------------------------------------------+

# openstack endpoint create --region RegionOne \
  object-store admin http://controller:8080/v1
```

```
+----------------+----------------------------------+
| Field          | Value                            |
+----------------+----------------------------------+
| enabled        | True                             |
| id             | ebb72cd6851d4defabc0b9d71cdca69b |
| interface      | admin                            |
| region         | RegionOne                        |
| region_id      | RegionOne                        |
| service_id     | 75ef509da2c340499d454ae96a2c5c34 |
| service_name   | swift                            |
| service_type   | object-store                     |
| url            | http://controller:8080/v1        |
+----------------+----------------------------------+
```

## 17.1.2　安装和配置 Swift 对象存储服务组件

下面开始安装和配置 Swift 对象存储服务组件，由于版本的不同，可能一些配置文件的内容也不同，需要修改或者添加一些内容。省略号（...）代表配置文件中需要保留的默认配置信息。

（1）安装软件包。

```
# zypper install openstack-swift-proxy python-swiftclient \
  python-keystoneclient python-keystonemiddleware \
  python-xml memcached
```

 注意：

以上需安装的软件包可能在之前安装其他服务组件时已经被安装，如 MemCached。

（2）编辑/etc/swift/proxy-server.conf 文件，完成以下操作。

① 在[DEFAULT]项，配置 Swift 对象存储服务组件使用的端口、用户和配置路径。

```
[DEFAULT]
...
bind_port = 8080
user = swift
swift_dir = /etc/swift
```

② 在[pipeline:main]项，启用相关的模块。

```
[pipeline:main]
pipeline = catch_errors gatekeeper healthcheck proxy-logging cache
container_sync bulk ratelimit authtoken keystoneauth container-quotas
account-quotas slo dlo versioned_writes proxy-logging proxy-server
```

③ 在[app:proxy-server]项，启用自动账户创建。

```
[app:proxy-server]
use = egg:swift#proxy
...
account_autocreate = true
```

④ 在[filter:keystoneauth]项，配置操作用户角色。

```
[filter:keystoneauth]
use = egg:swift#keystoneauth
...
operator_roles = admin,user
```

⑤ 在[filter:authtoken]项，配置 Keystone 身份认证服务组件访问。

```
[filter:authtoken]
paste.filter_factory = keystonemiddleware.auth_token:filter_factory
...
auth_uri = http://controller:5000
auth_url = http://controller:35357
auth_plugin = password
project_domain_id = default
user_domain_id = default
project_name = service
username = swift
password = SWIFT_PASS
delay_auth_decision = true
```

注意：
注释或者删除[filter:authtoken]项中的其他内容。

⑥ 在[filter:cache]项，配置 MemCached 的访问路径。

```
[filter:cache]
use = egg:swift#memcache
...
memcache_servers = 127.0.0.1:11211
```

## 17.2 安装和配置（存储节点）

本节详细描述如何在存储节点上安装和配置 Swift 对象存储服务组件。Swift 对象存储服务组件要求至少有两个存储节点，支持横向扩展。在本例中采用两个存储节点，每个存储节点包含两个块存储设备，分别是/dev/sdb 和/dev/sdc。

Swift 对象存储具有的 xattr（extended attributes）特性使得它支持各种类型的文件系统，如 Ext4、Ext3；但是基于测试完整性、性能和可靠性考虑，这里选择 XFS 文件系统。以下操作步骤需要在每个存储节点上执行。

### 17.2.1 准备

在安装和配置之前，请务必准备好块存储设备。

（1）安装软件包。

```
# zypper install xfsprogs rsync
```

（2）格式化/dev/sdb 和/dev/sdc 块存储设备为 XFS。

```
# mkfs.xfs /dev/sdb
# mkfs.xfs /dev/sdc
```

（3）创建挂载点。

```
# mkdir -p /srv/node/sdb
# mkdir -p /srv/node/sdc
```

（4）编辑/etc/fstab 文件，添加以下内容。

```
/dev/sdb /srv/node/sdb xfs noatime,nodiratime,nobarrier,logbufs=8 0 2
```

```
/dev/sdc /srv/node/sdc xfs noatime,nodiratime,nobarrier,logbufs=8 0 2
```

(5) 挂载文件系统。

```
# mount /srv/node/sdb
# mount /srv/node/sdc
```

(6) 编辑/etc/rsyncd.conf 文件，添加以下内容。

```
uid = swift
gid = swift
log file = /var/log/rsyncd.log
pid file = /var/run/rsyncd.pid
address = MANAGEMENT_INTERFACE_IP_ADDRESS

[account]
max connections = 2
path = /srv/node/
read only = false
lock file = /var/lock/account.lock

[container]
max connections = 2
path = /srv/node/
read only = false
lock file = /var/lock/container.lock

[object]
max connections = 2
path = /srv/node/
read only = false
lock file = /var/lock/object.lock
```

替换 MANAGEMENT_INTERFACE_IP_ADDRESS 为存储节点的管理/数据 IP 地址。

注意：

rsyncd 服务在使用时不需要认证和验证，所以可以使用单独的私有网络运行。

(7) 启动 rsyncd 服务并设置为开机自启动。

```
# systemctl enable rsyncd.service
# systemctl start rsyncd.service
```

## 17.2.2 安装和配置 Swift 对象存储服务组件

下面开始安装和配置 Swift 对象存储服务组件，由于版本的不同，可能一些配置文件的内容也不同，需要修改或者添加一些内容。省略号（…）代表配置文件中需要保留的默认配置信息。

（1）安装软件包。

```
# zypper install openstack-swift-account \
  openstack-swift-container openstack-swift-object python-xml
```

（2）编辑/etc/swift/account-server.conf 文件，完成以下操作。

① 在[DEFAULT]项，配置 Account 使用的 IP 地址、端口、用户、配置路径和挂载点。

```
[DEFAULT]
...
bind_ip = MANAGEMENT_INTERFACE_IP_ADDRESS
bind_port = 6002
user = swift
swift_dir = /etc/swift
devices = /srv/node
mount_check = true
```

替换 MANAGEMENT_INTERFACE_IP_ADDRESS 为存储节点的管理/数据 IP 地址。

② 在[pipeline:main]项，启用相关的功能模块。

```
[pipeline:main]
pipeline = healthcheck recon account-server
```

③ 在[filter:recon]项，配置 rcon 缓存目录。

```
[filter:recon]
use = egg:swift#recon
...
recon_cache_path = /var/cache/swift
```

（3）编辑/etc/swift/container-server.conf 文件，完成以下操作。

① 在[DEFAULT]项，配置 Container 使用的 IP 地址、端口、用户、配置路径和挂载点。

```
[DEFAULT]
...
bind_ip = MANAGEMENT_INTERFACE_IP_ADDRESS
bind_port = 6001
user = swift
swift_dir = /etc/swift
devices = /srv/node
mount_check = true
```

替换 MANAGEMENT_INTERFACE_IP_ADDRESS 为存储节点的管理/数据 IP 地址。

② 在[pipeline:main]项，启用相关的功能模块。

```
[pipeline:main]
pipeline = healthcheck recon container-server
```

③ 在[filter:recon]项，配置 rcon 缓存目录。

```
[filter:recon]
use = egg:swift#recon
...
recon_cache_path = /var/cache/swift
```

(4) 编辑/etc/swift/object-server.conf 文件，完成以下操作。

① 在[DEFAULT]项，配置 Object 使用的 IP 地址、端口、用户、配置路径和挂载点。

```
[DEFAULT]
...
bind_ip = MANAGEMENT_INTERFACE_IP_ADDRESS
bind_port = 6000
user = swift
swift_dir = /etc/swift
devices = /srv/node
mount_check = true
```

替换 MANAGEMENT_INTERFACE_IP_ADDRESS 为存储节点的管理/数据 IP 地址。

② 在[pipeline:main]项，启用相关的功能模块。

```
[pipeline:main]
pipeline = healthcheck recon object-server
```

③ 在[filter:recon]项，配置 rcon 缓存目录和 lock 路径。

```
[filter:recon]
use = egg:swift#recon
...
recon_cache_path = /var/cache/swift
recon_lock_path = /var/lock
```

（5）设置挂载点的属主。

```
# chown -R root:swift /srv/node
```

（6）创建 rcon 目录并设置属主。

```
# mkdir -p /var/cache/swift
# chown -R root:swift /var/cache/swift
```

## 17.3　创建和分发 Ring

本节详细描述如何创建第一个 Account、Container 和 Object Ring，创建完成后会产生配置文件，存储节点会根据这些配置文件确定和部署存储架构。本例设置每个 Region 和 Zone 最多含有 1024 个 Partition，每个 Object 有 3 份副本，每个 Partition 在 1 小时内不会挪动多于一次。对于 Swift 对象存储服务组件，Partition 指的是目录而不是普通意义的分区表。以下操作均在控制节点上进行。

### 17.3.1　创建用户 Ring

Account 服务器使用 Account Ring 维护多个 Container。

（1）创建/etc/swift/account.builder 文件。

```
# swift-ring-builder account.builder create 10 3 1
```

（2）为 Account Ring 添加存储节点。

```
# swift-ring-builder account.builder add --region 1 --zone 1 \
--ip STORAGE_NODE_MANAGEMENT_INTERFACE_IP_ADDRESS  --port 6002 \
  --device DEVICE_NAME --weight DEVICE_WEIGHT
```

替换 STORAGE_NODE_MANAGEMENT_INTERFACE_IP_ADDRESS 为存储节点的管理/数据 IP 地址，替换 DEVICE_NAME 为存储节点的块存储设备。

```
# swift-ring-builder account.builder add \
  --region 1 --zone 1 --ip 172.168.1.14 --port 6002 --device sdb \
  --weight 100
```

在每个存储节点上执行上述命令，根据本例的配置，输入如下内容。

```
# swift-ring-builder account.builder add \
  --region 1 --zone 1 --ip 172.168.1.14 --port 6002 --device sdb \
  --weight 100
Device  d0r1z1-172.168.1.14:6002R172.168.1.14:6002/sdb_""  with  100.0 weight got id 0
# swift-ring-builder account.builder add \
  --region 1 --zone 2 --ip 172.168.1.14 --port 6002 --device sdc \
  --weight 100
Device  d1r1z2-172.168.1.14:6002R172.168.1.14:6002/sdc_""  with  100.0 weight got id 1
# swift-ring-builder account.builder add \
  --region 1 --zone 3 --ip 172.168.1.15 --port 6002 --device sdb \
  --weight 100
Device  d2r1z3-172.168.1.15:6002R172.168.1.15:6002/sdb_""  with  100.0 weight got id 2
# swift-ring-builder account.builder add \
  --region 1 --zone 4 --ip 172.168.1.15 --port 6002 --device sdc \
  --weight 100
Device  d3r1z4-172.168.1.15:6002R172.168.1.15:6002/sdc_""  with  100.0 weight got id 3
```

（3）确认 Ring 内容。

```
# swift-ring-builder account.builder
account.builder, build version 4
1024 partitions, 3.000000 replicas, 1 regions, 4 zones, 4 devices, 100.00 balance, 0.00 dispersion
The minimum number of hours before a partition can be reassigned is 1
The overload factor is 0.00% (0.000000)
```

(4) 重平衡 Ring。

```
# swift-ring-builder account.builder rebalance
Reassigned 1024 (100.00%) partitions. Balance is now 0.00.  Dispersion is now 0.00
```

## 17.3.2 创建 Container Ring

Container 服务器使用 Container Ring 维护多个 Object。但是 Container 并不能确定 Object 的位置。

(1) 创建 /etc/swift/container.builder 文件。

```
# swift-ring-builder container.builder create 10 3 1
```

(2) 为 Container Ring 添加存储节点。

```
# swift-ring-builder container.builder add --region 1 --zone 1 \
--ip STORAGE_NODE_MANAGEMENT_INTERFACE_IP_ADDRESS --port 6001 \
  --device DEVICE_NAME --weight DEVICE_WEIGHT
```

替换 STORAGE_NODE_MANAGEMENT_INTERFACE_IP_ADDRESS 为存储节点的管理/数据 IP 地址，替换 DEVICE_NAME 为存储节点的块存储设备。

```
# swift-ring-builder container.builder add \
  --region 1 --zone 1 --ip 172.168.1.14 --port 6001 --device sdb \
--weight 100
```

在每个存储节点上执行上述命令，根据本例的配置，输入如下内容。

```
# swift-ring-builder container.builder add \
  --region 1 --zone 1 --ip 172.168.1.14 --port 6001 --device sdb \
--weight 100
Device    d0r1z1-172.168.1.14:6001R172.168.1.14:6001/sdb_""  with  100.0 weight got id 0
# swift-ring-builder container.builder add \
  --region 1 --zone 2 --ip 172.168.1.14 --port 6001 --device sdc \
--weight 100
Device    d1r1z2-172.168.1.14:6001R172.168.1.14:6001/sdc_""  with  100.0 weight got id 1
```

```
# swift-ring-builder container.builder add \
  --region 1 --zone 3 --ip 172.168.1.15 --port 6001 --device sdb \
--weight 100
Device  d2r1z3-172.168.1.15:6001R172.168.1.15:6001/sdb_""  with  100.0
weight got id 2
# swift-ring-builder container.builder add \
  --region 1 --zone 4 --ip 172.168.1.15 --port 6001 --device sdc \
--weight 100
Device  d3r1z4-172.168.1.15:6001R172.168.1.15:6001/sdc_""  with  100.0
weight got id 3
```

(3) 确认 Ring 内容。

```
# swift-ring-builder container.builder
container.builder, build version 4
1024 partitions, 3.000000 replicas, 1 regions, 4 zones, 4 devices, 100.00 balance, 0.00 dispersion
The minimum number of hours before a partition can be reassigned is 1
The overload factor is 0.00% (0.000000)
```

(4) 重平衡 Ring。

```
# swift-ring-builder container.builder rebalance
Reassigned 1024 (100.00%) partitions. Balance is now 0.00.  Dispersion is now 0.00
```

## 17.3.3 创建 Object Ring

Object 服务器使用 Object Ring 维护 Object 的位置信息。

(1) 创建/etc/swift/ object.builder 文件。

```
# swift-ring-builder object.builder create 10 3 1
```

(2) 为 Object Ring 添加存储节点。

```
# swift-ring-builder object.builder add --region 1 --zone 1 \
--ip STORAGE_NODE_MANAGEMENT_INTERFACE_IP_ADDRESS --port 6000 \
  --device DEVICE_NAME --weight DEVICE_WEIGHT
```

替换 STORAGE_NODE_MANAGEMENT_INTERFACE_IP_ADDRESS 为存储节点的管理/数据 IP 地址，替换 DEVICE_NAME 为存储节点的块存储设备。

```
# swift-ring-builder object.builder add \
  --region 1 --zone 1 --ip 172.168.1.14 --port 6001 --device sdb \
--weight 100
```

在每个存储节点上执行上述命令，根据本例的配置，输入如下内容。

```
# swift-ring-builder object.builder add \
  --region 1 --zone 1 --ip 172.168.1.14 --port 6000 --device sdb \
--weight 100
Device   d0r1z1-172.168.1.14:6000R172.168.1.14:6000/sdb_""   with   100.0 weight got id 0
# swift-ring-builder object.builder add \
  --region 1 --zone 2 --ip 172.168.1.14 --port 6000 --device sdc \
--weight 100
Device   d1r1z2-172.168.1.14:6000R172.168.1.14:6000/sdc_""   with   100.0 weight got id 1
# swift-ring-builder object.builder add \
  --region 1 --zone 3 --ip 172.168.1.15 --port 6000 --device sdb \
--weight 100
Device   d2r1z3-172.168.1.15:6000R172.168.1.15:6000/sdb_""   with   100.0 weight got id 2
# swift-ring-builder object.builder add \
  --region 1 --zone 4 --ip 172.168.1.15 --port 6000 --device sdc \
--weight 100
Device   d3r1z4-172.168.1.15:6000R172.168.1.15:6000/sdc_""   with   100.0 weight got id 3
```

（3）确认 Ring 内容。

```
# swift-ring-builder object.builder
object.builder, build version 4
1024 partitions, 3.000000 replicas, 1 regions, 4 zones, 4 devices, 100.00 balance, 0.00 dispersion
The minimum number of hours before a partition can be reassigned is 1
The overload factor is 0.00% (0.000000)
```

(4) 重平衡 Ring。

```
# swift-ring-builder object.builder rebalance
Reassigned 1024 (100.00%) partitions. Balance is now 0.00. Dispersion is now 0.00
```

## 17.3.4 分发 Ring 配置文件

复制控制节点上的/etc/swift/account.ring.gz、/etc/swift/container.ring.gz 和/etc/swift/object.ring.gz 文件到每个存储节点的/etc/swift 目录下，后续添加新的存储节点时，每个运行 Proxy 服务的存储节点上都需要执行此操作。

## 17.4 安装完成

（1）在控制节点上编辑/etc/swift/swift.conf 文件，完成以下操作。

① 在[swift-hash]项，配置 Hash 路径的前缀和后缀。

```
[swift-hash]
...
swift_hash_path_suffix = HASH_PATH_SUFFIX
swift_hash_path_prefix = HASH_PATH_PREFIX
```

使用随机、唯一的字符串替换 HASH_PATH_SUFFIX 和 HASH_PATH_PREFIX 字段。

 警告：
该字符串必须是安全的，并且不能修改或者丢弃。

② 在[storage-policy:0]项，配置默认的存储策略。

```
[storage-policy:0]
...
name = Policy-0
default = yes
```

（2）分发 swift.conf 文件到每个存储节点的/etc/swift 目录下，后续添加新的存储节点时，每个运行 Proxy 服务的存储节点上都需要执行此操作。

（3）在所有节点上，确认/etc/swift 目录的属主正确。

```
# chown -R root:swift /etc/swift
```

（4）在控制节点和运行 Proxy 服务的存储节点上，启动 Proxy 服务及相关服务，并设置开机自启动。

```
# systemctl enable openstack-swift-proxy.service memcached.service
# systemctl start openstack-swift-proxy.service memcached.service
```

（5）在存储节点上，启动 Swift 对象存储服务并设置为开机自启动。

```
# systemctl enable openstack-swift-account.service \
openstack-swift-account-auditor.service \
openstack-swift-account-reaper.service \
openstack-swift-account-replicator.service
# systemctl start openstack-swift-account.service \
  openstack-swift-account-auditor.service \
  openstack-swift-account-reaper.service \
openstack-swift-account-replicator.service
# systemctl enable openstack-swift-container.service \
  openstack-swift-container-auditor.service \
  openstack-swift-container-replicator.service \
openstack-swift-container-updater.service
# systemctl start openstack-swift-container.service \
  openstack-swift-container-auditor.service \
  openstack-swift-container-replicator.service \
openstack-swift-container-updater.service
# systemctl enable openstack-swift-object.service \
  openstack-swift-object-auditor.service \
  openstack-swift-object-replicator.service \
openstack-swift-object-updater.service
# systemctl start openstack-swift-object.service \
  openstack-swift-object-auditor.service \
  openstack-swift-object-replicator.service \
openstack-swift-object-updater.service
```

## 17.5 验证

下面检查一下配置完成的 Swift 对象存储服务组件。所有的操作均在控制节点上进行。

(1) 添加环境变量条目。

```
# echo "export OS_AUTH_VERSION=3" \
  | tee -a admin-openrc.sh demo-openrc.sh
```

(2) 加载环境变量。

```
# source demo-openrc.sh
```

(3) 检查 Swift 对象存储服务运行状态。

```
# swift stat
                        Account: AUTH_ed0b60bf607743088218b0a533d5943f
                     Containers: 0
                        Objects: 0
                          Bytes: 0
  Containers in policy "policy-0": 0
     Objects in policy "policy-0": 0
       Bytes in policy "policy-0": 0
      X-Account-Project-Domain-Id: default
                    X-Timestamp: 1444143887.71539
                     X-Trans-Id: tx1396aeaf17254e94beb34-0056143bde
                   Content-Type: text/plain; charset=utf-8
                   Accept-Ranges: bytes
```

(4) 上传文件进行测试。

```
# swift upload container1 FILE
FILE
```

替换 FILE 字段为上传文件的名字。

(5) 列出所有的 Container。

```
# swift list
container1
```

(6)下载文件进行测试。

```
# swift download container1 FILE
FILE [auth 0.141s, headers 0.278s, total 0.278s, 0.05 MB/s]
```

替换 FILE 字段为 container1 中存在的文件名。

# 第 18 章 编排（Heat）服务安装配置

OpenStack 项目中的 Heat 编排服务组件是编排资源的一个工具，它能够生成一个模板，该模板通过资源、参数、输入、彼此的约束和依赖等参数描述被执行的任务。Heat 编排服务组件使用 YAML 编写的 Heat Orchestration Template（HOT）模板。YAML 是轻量化的数据描述语言，语法比 XML 简单很多，类似 Pyson 和 Ruby。

Heat 编排服务组件可以通过 CLI 和 RESTful 两种方式进行访问，并且集成到 Horizon 仪表板服务组件中，可以通过 Web 界面对其进行操作。Heat 编排服务组件提供了兼容 AWS CloudFormation 模板的 API。

## 18.1 安装和配置

本节详细描述如何在控制节点上安装和配置 Heat 编排服务组件。

### 18.1.1 准备

在安装和配置 Heat 编排服务组件之前，首先创建数据库、服务证书（service credential）和 API endpoint。Heat 编排服务组件需要在 Keystone 身份认证服务中添加信息。

（1）创建数据库。

① 在操作系统终端连接数据库。

```
# mysql -u root -p
```

因为我们使用 root 用户登录，所以在提示输入密码时，按回车键即可建立连接。

② 创建 Heat 数据库。

```
MariaDB [(none)]> CREATE DATABASE heat;
```

③ Heat 数据库的访问权限设置。

```
GRANT ALL PRIVILEGES ON heat.* TO 'heat'@'localhost' \
  IDENTIFIED BY 'HEAT_DBPASS';
GRANT ALL PRIVILEGES ON heat.* TO 'heat'@'%' \
  IDENTIFIED BY 'HEAT_DBPASS';
```

替换 HEAT_DBPASS 为合适的密码。

④ 退出数据库。

（2）加载 admin user 的环境变量。

```
# source admin-openrc.sh
```

（3）创建 Orchestration 服务凭据。

① 创建 Heat 用户。

```
# openstack user create --domain default --password-prompt heat
User Password:
Repeat User Password:
```

| Field | Value |
| --- | --- |
| domain_id | default |
| enabled | True |
| id | ca2e175b851943349be29a328cc5e360 |
| name | heat |

② 将 admin role 赋予 heat user 和 service project。

# openstack role add --project service --user cinder admin

**注意：**
此命令执行完成后，没有输出。

③ 创建 Heat 和 heat-cfn service entity。

```
# openstack service create --name heat \
  --description "Orchestration" orchestration
```

```
+-------------+----------------------------------+
| Field       | Value                            |
+-------------+----------------------------------+
| description | Orchestration                    |
| enabled     | True                             |
| id          | 727841c6f5df4773baa4e8a5ae7d72eb |
| name        | heat                             |
| type        | orchestration                    |
+-------------+----------------------------------+
```

```
# openstack service create --name heat-cfn \
  --description "Orchestration" cloudformation
```

```
+-------------+----------------------------------+
| Field       | Value                            |
+-------------+----------------------------------+
| description | Orchestration                    |
| enabled     | True                             |
| id          | c42cede91a4e47c3b10c8aedc8d890c6 |
| name        | heat-cfn                         |
| type        | cloudformation                   |
+-------------+----------------------------------+
```

(4) 创建 Heat 编排服务组件的 API endpoint。

```
# openstack endpoint create --region RegionOne \
  orchestration public http://controller:8004/v1/%\(tenant_id\)s
```

```
+----------------+------------------------------------------+
| Field          | Value                                    |
+----------------+------------------------------------------+
| enabled        | True                                     |
| id             | 3f4dab34624e4be7b000265f25049609         |
| interface      | public                                   |
| region         | RegionOne                                |
| region_id      | RegionOne                                |
| service_id     | 727841c6f5df4773baa4e8a5ae7d72eb         |
| service_name   | heat                                     |
| service_type   | orchestration                            |
| url            | http://controller:8004/v1/%(tenant_id)s  |
+----------------+------------------------------------------+
```

```
# openstack endpoint create --region RegionOne \
  orchestration internal http://controller:8004/v1/%\(tenant_id\)s
```

```
+----------------+------------------------------------------+
| Field          | Value                                    |
+----------------+------------------------------------------+
| enabled        | True                                     |
| id             | 9489f78e958e45cc85570fec7e836d98         |
| interface      | internal                                 |
| region         | RegionOne                                |
| region_id      | RegionOne                                |
| service_id     | 727841c6f5df4773baa4e8a5ae7d72eb         |
| service_name   | heat                                     |
| service_type   | orchestration                            |
| url            | http://controller:8004/v1/%(tenant_id)s  |
+----------------+------------------------------------------+
```

```
# openstack endpoint create --region RegionOne \
  orchestration admin http://controller:8004/v1/%\(tenant_id\)s
```

```
+----------------+------------------------------------------+
| Field          | Value                                    |
+----------------+------------------------------------------+
| enabled        | True                                     |
| id             | 76091559514b40c6b7b38dde790efe99         |
| interface      | admin                                    |
| region         | RegionOne                                |
| region_id      | RegionOne                                |
| service_id     | 727841c6f5df4773baa4e8a5ae7d72eb         |
| service_name   | heat                                     |
| service_type   | orchestration                            |
| url            | http://controller:8004/v1/%(tenant_id)s  |
+----------------+------------------------------------------+
```

```
# openstack endpoint create --region RegionOne \
  cloudformation public http://controller:8000/v1
```

```
+-------------+-----------------------------------+
| Field       | Value                             |
+-------------+-----------------------------------+
| enabled     | True                              |
| id          | b3ea082e019c4024842bf0a80555052c  |
| interface   | public                            |
| region      | RegionOne                         |
| region_id   | RegionOne                         |
| service_id  | c42cede91a4e47c3b10c8aedc8d890c6  |
| service_name| heat-cfn                          |
| service_type| cloudformation                    |
| url         | http://controller:8000/v1         |
+-------------+-----------------------------------+
```

```
# openstack endpoint create --region RegionOne \
  cloudformation internal http://controller:8000/v1
```

```
+-------------+-----------------------------------+
| Field       | Value                             |
+-------------+-----------------------------------+
| enabled     | True                              |
| id          | 169df4368cdc435b8b115a9cb084044e  |
| interface   | internal                          |
| region      | RegionOne                         |
| region_id   | RegionOne                         |
| service_id  | c42cede91a4e47c3b10c8aedc8d890c6  |
| service_name| heat-cfn                          |
| service_type| cloudformation                    |
| url         | http://controller:8000/v1         |
+-------------+-----------------------------------+
```

```
# openstack endpoint create --region RegionOne \
  cloudformation admin http://controller:8000/v1
```

```
+-------------+-----------------------------------+
| Field       | Value                             |
+-------------+-----------------------------------+
| enabled     | True                              |
| id          | 3d3edcd61eb343c1bbd629aa041ff88b  |
| interface   | internal                          |
| region      | RegionOne                         |
| region_id   | RegionOne                         |
| service_id  | c42cede91a4e47c3b10c8aedc8d890c6  |
| service_name| heat-cfn                          |
| service_type| cloudformation                    |
| url         | http://controller:8000/v1         |
+-------------+-----------------------------------+
```

（5）Orchestration 服务需要在 Identity 服务中添加信息，用以操作 Stack。

① 创建 Heat 域。

```
# openstack domain create --description "Stack projects and users" heat
```

```
+-------------+----------------------------------+
| Field       | Value                            |
+-------------+----------------------------------+
| description | Stack projects and users         |
| enabled     | True                             |
| id          | 0f4d1bd326f2454dacc72157ba328a47 |
| name        | heat                             |
+-------------+----------------------------------+
```

② 创建 heat_domain_admin 用户，管理 Heat 域中的 project 和 user。

```
# openstack user create --domain heat --password-prompt heat_domain_admin
User Password:
Repeat User Password:
```

```
+-----------+----------------------------------+
| Field     | Value                            |
+-----------+----------------------------------+
| domain_id | 0f4d1bd326f2454dacc72157ba328a47 |
| enabled   | True                             |
| id        | b7bd1abfbcf64478b47a0f13cd4d970a |
| name      | heat_domain_admin                |
+-----------+----------------------------------+
```

③ 将 admin role 赋予 heat_domain_admin 用户。

```
# openstack role add --domain heat --user heat_domain_admin admin
```

**注意：**
此命令执行完成后，没有输出。

④ 创建 heat_stack_owner role。

```
# openstack role create heat_stack_owner
```

```
+-------+----------------------------------+
| Field | Value                            |
+-------+----------------------------------+
| id    | 15e34f0c4fed4e68b3246275883c8630 |
| name  | heat_stack_owner                 |
+-------+----------------------------------+
```

⑤ 将 heat_stack_owner role 赋予 demo project 和 user，使 demo user 可以管理 Stack。

```
# openstack role add --project demo --user demo heat_stack_owner
```

注意：

此命令执行完成后，没有输出。

注意：

必须将 heat_stack_owner role 赋予每个管理 Stack 的用户。

⑥ 创建 heat_stack_user role。

```
# openstack role create heat_stack_user
```

注意：

Heat 编排服务组件会自动将 heat_stack_user role 赋予在部署 Stack 时创建的用户。为了避免冲突，不要将 heat_stack_owner role 赋予用户。

## 18.1.2　安装和配置 Heat 编排服务组件

下面开始安装和配置 Heat 编排服务组件，由于版本的不同，可能一些配置文件的内容也不同，需要修改或者添加一些内容。省略号（...）代表配置文件中需要保留的默认配置信息。

（1）安装软件包。

```
# zypper install openstack-heat-api openstack-heat-api-cfn \
  openstack-heat-engine python-heatclient
```

（2）编辑/etc/cinder/cinder.conf 文件，完成以下操作。

① 在[database]项，配置数据库连接。

```
[database]
...
connection = mysql+pymysql://heat:HEAT_DBPASS@controller/heat
```

使用已经定义的 Heat 数据库密码替换 HEAT_DBPASS。

② 在[DEFAULT]和[oslo_messaging_rabbit]项，配置 RabbitMQ 消息队列连接。

```
[DEFAULT]
...
rpc_backend = rabbit

[oslo_messaging_rabbit]
...
rabbit_host = controller
rabbit_userid = openstack
rabbit_password = RABBIT_PASS
```

使用已经在 RabbitMQ 中定义的 OpenStack 用户密码替换 RABBIT_PASS。

③ 在 [keystone_authtoken]、[trustee]、[clients_keystone] 和 [ec2authtoken] 项，配置 Keystone 身份认证服务组件连接。

```
[keystone_authtoken]
...
auth_uri = http://controller:5000
auth_url = http://controller:35357
auth_plugin = password
project_domain_id = default
user_domain_id = default
project_name = service
username = heat
password = HEAT_PASS

[trustee]
...
auth_plugin = password
auth_url = http://controller:35357
username = heat
password = HEAT_PASS
```

```
user_domain_id = default

[clients_keystone]
...
auth_uri = http://controller:5000

[ec2authtoken]
...
auth_uri = http://controller:5000
```

使用已经定义的 Heat 用户密码替换 HEAT_PASS。

④ 在[DEFAULT]项，配置 metadata 和 waitcondition 访问地址。

```
[DEFAULT]
...
heat_metadata_server_url = http://controller:8000
heat_waitcondition_server_url = http://controller:8000/v1/waitcondition
```

⑤ 在[DEFAULT]项，配置 Stack 域和管理凭据。

```
[DEFAULT]
...
stack_domain_admin = heat_domain_admin
stack_domain_admin_password = HEAT_DOMAIN_PASS
stack_user_domain_name = heat
```

使用已经定义的 heat_domain_admin 用户密码替换 HEAT_DOMAIN_PASS。

⑥ 打开日志记录功能，方便问题跟踪和解决。

```
[DEFAULT]
...
verbose = True
```

（3）同步信息到 Orchestration 数据库。

```
# su -s /bin/sh -c "heat-manage db_sync" heat
```

## 18.1.3 安装完成

启动 Heat 编排服务并设置为开机自启动。

```
# systemctl enable openstack-heat-api.service \
  openstack-heat-api-cfn.service openstack-heat-engine.service
# systemctl start openstack-heat-api.service \
  openstack-heat-api-cfn.service openstack-heat-engine.service
```

## 18.2 验证

下面检查一下配置完成的 Heat 编排服务组件。所有的操作均在控制节点上进行。

(1) 加载环境变量。

```
# source admin-openrc.sh
```

(2) 列出 Heat 编排服务组件进程是否成功启动和注册。

```
# heat service-list
+------------+-------------+--------------------------------------+------------+--------+
| hostname   | binary      | engine_id                            | host       | topic  |
+------------+-------------+--------------------------------------+------------+--------+
| controller | heat-engine | 3e85d1ab-a543-41aa-aa97-378c381fb958 | controller | engine |
| controller | heat-engine | 45dbdcf6-5660-4d5f-973a-c4fc819da678 | controller | engine |
| controller | heat-engine | 51162b63-ecb8-4c6c-98c6-993af899c4f7 | controller | engine |
| controller | heat-engine | 8d7edc6d-77a6-460d-bd2a-984d76954646 | controller | engine |
+------------+-------------+--------------------------------------+------------+--------+
```

# 第 19 章 计量（Ceilometer）服务安装配置

Ceilometer 计量服务组件提供了统计、计费等功能，总结如下：

- 循环收集有关 OpenStack 项目中各服务组件的计量数据。
- 通过监控各种通知收集事件和计量数据。
- 发布数据，包括数据存储和消息队列。
- 收集到的数据超出预定义的阈值，发出警告。

## 19.1 安装和配置

本节详细描述如何在控制节点上安装和配置 Ceilometer 计量服务组件。Ceilometer 计量服务组件可以收集 OpenStack 项目中各服务组件的数据并选择性地触发报警信息。

### 19.1.1 准备

在安装和配置 Ceilometer 计量服务组件之前，首先创建数据库、服务证书（service

credential）和 API endpoint。但是，Ceilometer 计量服务组件和其他服务组件不同，需要创建 NoSQL 数据库 MongoDB。

（1）创建 Ceilometer 数据库。

```
# mongo --host controller --eval '
  db = db.getSiblingDB("ceilometer");
  db.addUser({user: "ceilometer",
  pwd: "CEILOMETER_DBPASS",
  roles: [ "readWrite", "dbAdmin" ]})'

MongoDB shell version: 2.4.x
connecting to: controller:27017/test
{
 "user" : "ceilometer",
 "pwd" : "72f25aeee7ad4be52437d7cd3fc60f6f",
 "roles" : [
  "readWrite",
  "dbAdmin"
 ],
 "_id" : ObjectId("5489c22270d7fad1ba631dc3")
}
```

替换 CEILOMETER_DBPASS 为合适的密码。

（2）加载 admin user 的环境变量。

```
# source admin-openrc.sh
```

（3）创建 Orchestration 服务凭据。

① 创建 Ceilometer 用户。

```
# openstack user create --domain default --password-prompt ceilometer
User Password:
Repeat User Password:
```

## 第 19 章 计量（Ceilometer）服务安装配置

```
+-----------+----------------------------------+
| Field     | Value                            |
+-----------+----------------------------------+
| domain_id | default                          |
| enabled   | True                             |
| id        | c859c96f57bd4989a8ea1a0b1d8ff7cd |
| name      | ceilometer                       |
+-----------+----------------------------------+
```

② 将 admin role 赋予 ceilometer user 和 service project。

```
# openstack role add --project service --user ceilometer admin
```

 **注意**：

此命令执行完成后，没有输出。

③ 创建 Ceilometer service entity。

```
# openstack service create --name ceilometer \
  --description "Telemetry" metering
```

```
+-------------+----------------------------------+
| Field       | Value                            |
+-------------+----------------------------------+
| description | Telemetry                        |
| enabled     | True                             |
| id          | 5fb7fd1bb2954fddb378d4031c28c0e4 |
| name        | ceilometer                       |
| type        | metering                         |
+-------------+----------------------------------+
```

（4）创建 Ceilometer 计量服务组件的 API endpoint。

```
# openstack endpoint create --region RegionOne \
  metering public http://controller:8777
```

```
+--------------+----------------------------------+
| Field        | Value                            |
+--------------+----------------------------------+
| enabled      | True                             |
| id           | b808b67b848d443e9eaaa5e5d796970c |
| interface    | public                           |
| region       | RegionOne                        |
| region_id    | RegionOne                        |
+--------------+----------------------------------+
```

```
| service_id   | 5fb7fd1bb2954fddb378d4031c28c0e4 |
| service_name | ceilometer                       |
| service_type | metering                         |
| url          | http://controller:8777           |
+--------------+----------------------------------+
```

```
# openstack endpoint create --region RegionOne \
  metering internal http://controller:8777
+--------------+----------------------------------+
| Field        | Value                            |
+--------------+----------------------------------+
| enabled      | True                             |
| id           | c7009b1c2ee54b71b771fa3d0ae4f948 |
| interface    | internal                         |
| region       | RegionOne                        |
| region_id    | RegionOne                        |
| service_id   | 5fb7fd1bb2954fddb378d4031c28c0e4 |
| service_name | ceilometer                       |
| service_type | metering                         |
| url          | http://controller:8777           |
+--------------+----------------------------------+
```

```
# openstack endpoint create --region RegionOne \
  metering admin http://controller:8777
+--------------+----------------------------------+
| Field        | Value                            |
+--------------+----------------------------------+
| enabled      | True                             |
| id           | b2c00566d0604551b5fe1540c699db3d |
| interface    | admin                            |
| region       | RegionOne                        |
| region_id    | RegionOne                        |
| service_id   | 5fb7fd1bb2954fddb378d4031c28c0e4 |
| service_name | ceilometer                       |
| service_type | metering                         |
| url          | http://controller:8777           |
+--------------+----------------------------------+
```

## 19.1.2 安装和配置 Ceilometer 计量服务组件

（1）安装软件包。

```
# zypper install openstack-ceilometer-api \
  openstack-ceilometer-collector \
  openstack-ceilometer-agent-notification \
  openstack-ceilometer-agent-central python-ceilometerclient \
  openstack-ceilometer-alarm-evaluator \
  openstack-ceilometer-alarm-notifier
```

（2）编辑/etc/ceilometer/ceilometer.conf 文件，完成以下操作。

① 在[database]项，配置数据库连接。

```
[database]
...
connection = mongodb://ceilometer:CEILOMETER_DBPASS@controller:27017/ceilometer
```

使用已经定义的 Ceilometer 数据库密码替换 CEILOMETER_DBPASS。注意，不能使用类似于":"、"/"、"+"和"@"等特殊字符。

② 在[DEFAULT]和[oslo_messaging_rabbit]项，配置 RabbitMQ 消息队列连接。

```
[DEFAULT]
...
rpc_backend = rabbit

[oslo_messaging_rabbit]
...
rabbit_host = controller
rabbit_userid = openstack
rabbit_password = RABBIT_PASS
```

使用已经在 RabbitMQ 中定义的 OpenStack 用户密码替换 RABBIT_PASS。

③ 在[DEFAULT]和[keystone_authtoken]项，配置 Keystone 身份认证服务组件连接。

```
[DEFAULT]
...
```

```
auth_strategy = keystone

[keystone_authtoken]
...
auth_uri = http://controller:5000
auth_url = http://controller:35357
auth_plugin = password
project_domain_id = default
user_domain_id = default
project_name = service
username = ceilometer
password = CEILOMETER_PASS
```

使用已经定义的 Ceilometer 用户密码替换 CEILOMETER_PASS。

④ 在[service_credentials]项，配置服务凭据。

```
[service_credentials]
...
os_auth_url = http://controller:5000/v2.0
os_username = ceilometer
os_tenant_name = service
os_password = CEILOMETER_PASS
os_endpoint_type = internalURL
os_region_name = RegionOne
```

使用已经定义的 Ceilometer 用户密码替换 CEILOMETER_PASS。

⑤ 在[collector]项，配置调度。

```
[collector]
...
dispatcher = database
```

⑥ 打开日志记录功能，方便问题跟踪和解决。

```
[DEFAULT]
...
verbose = True
```

## 19.1.3 安装完成

启动 Ceilometer 计量服务并设置为开机自启动。

```
# systemctl enable openstack-ceilometer-api.service \
  openstack-ceilometer-agent-notification.service \
  openstack-ceilometer-agent-central.service \
  openstack-ceilometer-collector.service \
  openstack-ceilometer-alarm-evaluator.service \
  openstack-ceilometer-alarm-notifier.service
# systemctl start openstack-ceilometer-api.service \
  openstack-ceilometer-agent-notification.service \
  openstack-ceilometer-agent-central.service \
  openstack-ceilometer-collector.service \
  openstack-ceilometer-alarm-evaluator.service \
  openstack-ceilometer-alarm-notifier.service
```

## 19.2 启用 Glance 镜像服务计量

Ceilometer 计量服务组件使用通知去收集 Glance 镜像服务的使用情况。以下操作均在控制节点上进行。

首先配置 Glance 镜像服务使用 Ceilometer 计量服务。

（1）编辑/etc/glance/glance-api.conf 和 /etc/glance/glance-registry.conf 文件，完成以下操作。

在[DEFAULT]和[oslo_messaging_rabbit]项，配置通知和 RabbitMQ 消息代理访问。

```
[DEFAULT]
...
notification_driver = messagingv2
rpc_backend = rabbit

[oslo_messaging_rabbit]
...
rabbit_host = controller
```

```
rabbit_userid = openstack
rabbit_password = RABBIT_PASS
```

使用已经在 RabbitMQ 中定义的 OpenStack 用户密码替换 RABBIT_PASS。

（2）重启 Image 服务。

```
# systemctl restart openstack-glance-api.service openstack-glance-registry.service
```

## 19.3 启用 Nova 计算服务计量

Ceilometer 计量服务组件使用通知和 agent 两种方式共同收集 Nova 计算服务的使用情况。以下操作均在计算节点上进行。

### 19.3.1 安装和配置 agent

（1）安装软件包。

```
# zypper install openstack-ceilometer-agent-compute
```

（2）编辑/etc/ceilometer/ceilometer.conf 文件，完成以下操作。

① 在[DEFAULT]和[oslo_messaging_rabbit]项，配置 RabbitMQ 消息队列连接。

```
[DEFAULT]
...
rpc_backend = rabbit

[oslo_messaging_rabbit]
...
rabbit_host = controller
rabbit_userid = openstack
rabbit_password = RABBIT_PASS
```

使用已经在 RabbitMQ 中定义的 OpenStack 用户密码替换 RABBIT_PASS。

② 在[DEFAULT]和[keystone_authtoken]项，配置 Keystone 身份认证服务组件连接。

```
[DEFAULT]
...
auth_strategy = keystone

[keystone_authtoken]
...
auth_uri = http://controller:5000
auth_url = http://controller:35357
auth_plugin = password
project_domain_id = default
user_domain_id = default
project_name = service
username = ceilometer
password = CEILOMETER_PASS
```

使用已经定义的 Ceilometer 用户密码替换 CEILOMETER_PASS。

③ 在[service_credentials]项，配置服务凭据。

```
[service_credentials]
...
os_auth_url = http://controller:5000/v2.0
os_username = ceilometer
os_tenant_name = service
os_password = CEILOMETER_PASS
os_endpoint_type = internalURL
os_region_name = RegionOne
```

使用已经定义的 Ceilometer 用户密码替换 CEILOMETER_PASS。

④ 打开日志记录功能，方便问题跟踪和解决。

```
[DEFAULT]
...
verbose = True
```

## 19.3.2 配置 Nova 计算服务使用 Ceilometer 计量服务

（1）编辑/etc/nova/nova.conf 文件，配置通知。

```
[DEFAULT]
...
instance_usage_audit = True
instance_usage_audit_period = hour
notify_on_state_change = vm_and_task_state
notification_driver = messagingv2
```

（2）启动 agent 并设置为开机自启动。

```
# systemctl enable openstack-ceilometer-agent-compute.service
# systemctl start openstack-ceilometer-agent-compute.service
```

（3）重启 Compute 服务。

```
# systemctl restart openstack-nova-compute.service
```

## 19.4 启用 Cinder 块存储服务计量

Ceilometer 计量服务组件使用通知去收集 Cinder 块存储服务的计量情况。以下操作均在控制节点和存储节点上进行。

首先配置 Cinder 块存储服务使用 Ceilometer 计量服务。

（1）编辑/etc/cinder/cinder.conf 文件，完成以下操作。

在[DEFAULT]项，配置通知。

```
[DEFAULT]
...
notification_driver = messagingv2
```

（2）重启控制节点上的 Cinder 块存储服务。

```
# systemctl restart openstack-cinder-api.service openstack-cinder-scheduler.service
```

（3）重启存储节点上的 Cinder 块存储服务。

```
# systemctl restart openstack-cinder-volume.service
```

（4）在存储节点上使用 cinder-volume-usage-audit 命令查看计量数据。

## 19.5 启用 Swift 对象存储服务计量

Telemetry 使用通知和循环查询两种方式共同收集 Swift 对象存储服务的计量情况。

### 19.5.1 准备

Ceilometer 计量服务组件使用 ResellerAdmin role 访问 Swift 对象存储服务。以下操作均在控制节点上进行。

（1）加载 admin user 的环境变量。

```
# source admin-openrc.sh
```

（2）创建 ResellerAdmin role。

```
# openstack role create ResellerAdmin
```

（3）将 ResellerAdmin role 赋予 ceilometer user 和 service project。

```
# openstack role add --project service --user ceilometer ResellerAdmin
```

注意：
此命令执行完成后，没有输出。

（4）安装软件包。

```
# zypper install python-ceilometermiddleware
```

## 19.5.2　配置 Swift 对象存储服务使用 Ceilometer 计量服务

以下操作在控制节点和运行 Proxy 服务的节点上进行。

（1）编辑/etc/swift/proxy-server.conf 文件，完成以下操作。

① 在[filter:keystoneauth]项，添加 ResellerAdmin role。

```
[filter:keystoneauth]
...
operator_roles = admin, user, ResellerAdmin
```

② 在[pipeline:main]项，添加 Ceilometer。

```
[pipeline:main]
pipeline = catch_errors gatekeeper healthcheck proxy-logging cache
container_sync bulk ratelimit authtoken keystoneauth container-quotas
account-quotas slo dlo versioned_writes proxy-logging ceilometer
proxy-server
```

③ 在[filter:ceilometer]项，配置通知。

```
[filter:ceilometer]
paste.filter_factory = ceilometermiddleware.swift:filter_factory
...
control_exchange = swift
url = rabbit://openstack:RABBIT_PASS@controller:5672/
driver = messagingv2
topic = notifications
log_level = WARN
```

使用已经在 RabbitMQ 中定义的 OpenStack 用户密码替换 RABBIT_PASS。

（2）重启 Swift 对象存储 Proxy 服务。

```
# systemctl restart openstack-swift-proxy.service
```

## 19.6　验证

下面检查一下配置完成的 Ceilometer 计量服务组件。简单起见，以下仅包括 Glance 镜像服务的计量情况。所有的操作均在控制节点上进行。

（1）加载 admin user 的环境变量。

```
# source admin-openrc.sh
```

（2）列出可用的计量服务。

```
# ceilometer meter-list
+------------+-------+------+--------------------------------------+---------+------------+
| Name       | Type  | Unit | Resource ID                          | User ID | Project ID |
+------------+-------+------+--------------------------------------+---------+------------+
| image      | gauge | image| acafc7c0-40aa-4026-9673-b879898e1fc2 | None    | cf12a15... |
| image.size | gauge | B    | acafc7c0-40aa-4026-9673-b879898e1fc2 | None    | cf12a15... |
+------------+-------+------+--------------------------------------+---------+------------+
```

（3）应用 Image 服务下载 CirrOS 镜像。

```
# IMAGE_ID=$(glance image-list | grep 'cirros' | awk '{ print $2 }')
# glance image-download $IMAGE_ID > /tmp/cirros.img
```

（4）再次列出可用的计量服务。

```
# ceilometer meter-list
+----------------+-------+------+--------------------------------------+---------+------------+
| Name           | Type  | Unit | Resource ID                          | User ID | Project ID |
+----------------+-------+------+--------------------------------------+---------+------------+
| image          | gauge | image| acafc7c0-40aa-4026-9673-b879898e1fc2 | None    | cf12a15... |
| image.download | delta | B    | acafc7c0-40aa-4026-9673-b879898e1fc2 | None    | cf12a15... |
| image.serve    | delta | B    | acafc7c0-40aa-4026-9673-b879898e1fc2 | None    | cf12a15... |
| image.size     | gauge | B    | acafc7c0-40aa-4026-9673-b879898e1fc2 | None    | cf12a15... |
+----------------+-------+------+--------------------------------------+---------+------------+
```

（5）查询 image.download 的使用统计。

```
# ceilometer statistics -m image.download -p 60
+--------+---------------------+---------------------+------------+------------+------------+------------+-------+
| Period | Period Start        | Period End          | Max        | Min        | Avg        | Sum        | Count |
+--------+---------------------+---------------------+------------+------------+------------+------------+-------+
| 60     | 2015-04-21T12:21:45 | 2015-04-21T12:22:45 | 13200896.0 | 13200896.0 | 13200896.0 | 13200896.0 | 1     |
+--------+---------------------+---------------------+------------+------------+------------+------------+-------+
```

（6）删除之前下载的 CirrOS 镜像。

```
# rm /tmp/cirros.img
```

# 第 20 章
# 建立虚拟机实例测试

本章操作均在控制节点上进行,从以下 5 个部分进行实例展示。

- 创建虚拟网络。
- 创建 Key Pair。
- 创建 Security Group 规则。
- 创建虚拟机实例。
- 创建块存储。

## 20.1 创建虚拟网络

本书介绍的网络服务组件提供两种架构,可以根据具体情况进行选择。

### 20.1.1 架构一网络(Public Provider Network)

在创建虚拟机实例之前,首先需要一个合适的网络架构。Public Provider Network 为虚拟机实例提供了基于 Layer 2(桥接和交换网络信息)的虚拟网络,虚拟网络中的 DHCP 为虚拟机实例提供了 IP 地址。

# 第 20 章 建立虚拟机实例测试

**注意**：

以下操作和截图均是示例，需要根据实际情况进行调整。

架构一：Public Provider Network 全局视图，如下图所示。

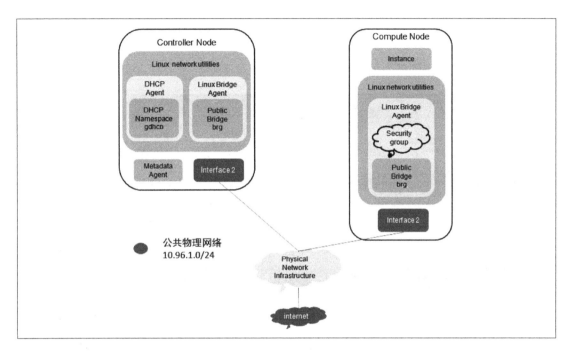

架构一：Public Provider Network 网络连接图，如下图所示。

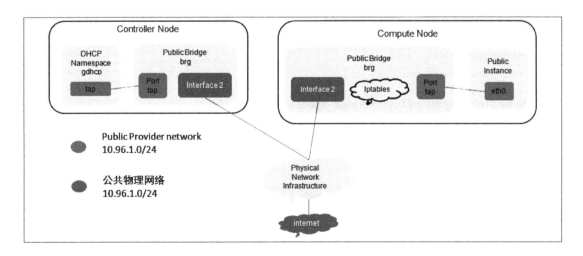

下面开始在控制节点上创建 Public Provider Network。

(1) 加载 admin user 的环境变量。

```
# source admin-openrc.sh
```

(2) 创建网络。

```
# neutron net-create public --shared --provider:physical_network public \
  --provider:network_type flat
Created a new network:
```

| Field | Value |
| --- | --- |
| admin_state_up | True |
| id | 37193c68-b70c-4153-b5cf-b6f19e9d8951 |
| mtu | 0 |
| name | public |
| port_security_enabled | True |
| provider:network_type | flat |
| provider:physical_network | public |
| provider:segmentation_id | |
| router:external | False |
| shared | True |
| status | ACTIVE |
| subnets | b51b9bdb-05d8-40c5-aa73-c70557bc46a9 |
| | a60f0452-ba3d-41bd-83d7-86a38c12f5b5 |
| tenant_id | efb33dccf25b40b399e92d6351fc7ea9 |

- --shared 参数：允许所有 project（租户）使用该虚拟网络。

- --provider:physical_network public 和--provider:network_type flat 参数：flat 类型的虚拟网络连接到 flat 类型的物理网络的 eth1 网络端口。该信息在配置文件中均已定义。

/etc/neutron/plugins/ml2/ml2_conf.ini 文件：

```
[ml2_type_flat]
flat_networks = public
```

/etc/neutron/plugins/ml2/linuxbridge_agent.ini 文件：

```
[linux_bridge]
physical_interface_mappings = public:eth1
```

(3) 创建子网。

```
# neutron subnet-create public PUBLIC_NETWORK_CIDR --name public \
  --allocation-pool start=START_IP_ADDRESS,end=END_IP_ADDRESS\
```

```
--dns-nameserver DNS_RESOLVER --gateway PUBLIC_NETWORK_GATEWAY
```

替换 PUBLIC_NETWORK_CIDR 为物理网络的 IP 地址范围，如 10.96.1.0/24。

替换 START_IP_ADDRESS 和 END_IP_ADDRESS 为子网的第一个 IP 地址和最后一个 IP 地址，该子网中不能存在已经被使用的 IP 地址。

替换 DNS_RESOLVER 为 DNS 解析服务器的地址，一般该地址可参考计算节点的 /etc/resolv.conf 中的内容条目。

替换 PUBLIC_NETWORK_GATEWAY 为物理网络的网关 IP 地址。

示例：公共网络的 IP 地址范围是 10.96.1.0/24，网关是 10.96.1.1，DHCP 服务器在 10.96.1.101～10.96.1.200 地址段内给虚拟机实例分配 IP 地址。

```
# neutron subnet-create public 10.96.1.0/24 --name public \
  --allocation-pool start=10.96.1.101,end=10.96.1.200 \
  --gateway 10.96.1.1
```

```
+-------------------+------------------------------------------------+
| Field             | Value                                          |
+-------------------+------------------------------------------------+
| allocation_pools  | {"start": "10.96.1.101", "end": "10.96.1.200"} |
| cidr              | 10.96.1.0/24                                   |
| dns_nameservers   |                                                |
| enable_dhcp       | True                                           |
| gateway_ip        | 10.96.1.1                                      |
| host_routes       |                                                |
| id                | b51b9bdb-05d8-40c5-aa73-c70557bc46a9           |
| ip_version        | 4                                              |
| ipv6_address_mode |                                                |
| ipv6_ra_mode      |                                                |
| name              | public                                         |
| network_id        | 37193c68-b70c-4153-b5cf-b6f19e9d8951           |
| subnetpool_id     |                                                |
| tenant_id         | efb33dccf25b40b399e92d6351fc7ea9               |
+-------------------+------------------------------------------------+
```

## 20.1.2 架构二网络（Private Project Network）

网络架构二是 Private Project Network，该虚拟网络通过 Layer 3（路由）和 NAT 功能连接到物理网络中，DHCP 服务为虚拟机实例提供 IP 地址。在该虚拟网络中的虚拟机实例可以直接连接外网，如 Internet。但是，从外网访问 Private Project Network 中的虚拟机实例需要 Floating IP 地址。

注意：

　　Floating IP 地址就是一个普通的 IP 地址，当从租户外面连接租户内的虚拟机实例时，需要使用此 IP 地址。虚拟机实例会始终拥有这个 IP 地址，无论是重启还是维护。

 注意：

　　以下操作和截图均是示例，需要根据实际情况进行调整。

 警告：

　　在创建 Private Project Network 之前，必须首先创建 Public Provider Network。

架构二：Private Project Network 全局视图，如下图所示。

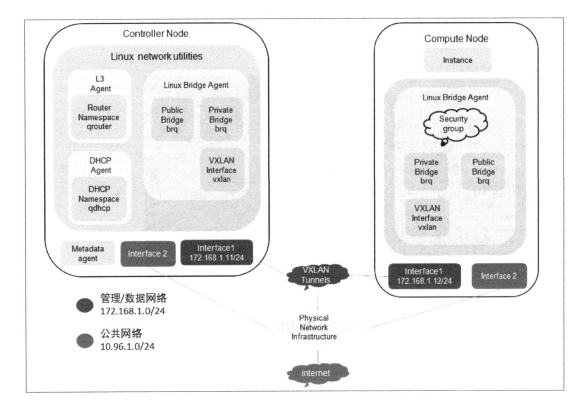

架构二：Private Project Network 网络连接图，如下图所示。

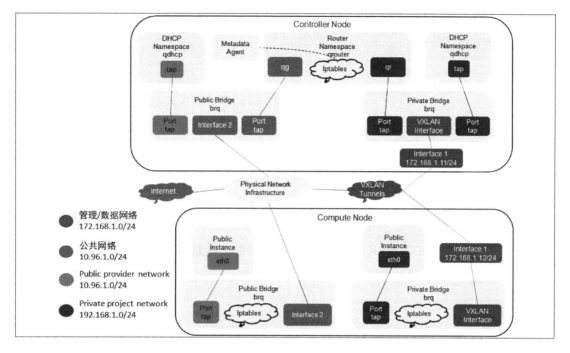

下面开始创建 Private Project Network。

（1）在控制节点上加载 demo user 的环境变量。

# source demo-openrc.sh

（2）创建网络。

```
# neutron net-create private
Created a new network:
+-----------------------+--------------------------------------+
| Field                 | Value                                |
+-----------------------+--------------------------------------+
| admin_state_up        | True                                 |
| id                    | 7c6f9b37-76b4-463e-98d8-27e5686ed083 |
| mtu                   | 0                                    |
| name                  | private                              |
| port_security_enabled | True                                 |
| router:external       | False                                |
| shared                | False                                |
| status                | ACTIVE                               |
| subnets               |                                      |
| tenant_id             | f5b2ccaa75ac413591f12fcaa096aa5c     |
+-----------------------+--------------------------------------+
```

非管理员用户使用该命令创建网络时，一般不支持手动添加任何参数，使用的默认参数已经在配置文件中进行定义。

ml2_conf.ini 文件：

```
[ml2]
tenant_network_types = vxlan

[ml2_type_vxlan]
vni_ranges = 1:1000
```

（3）基于此网络创建子网。

```
# neutron subnet-create private PRIVATE_NETWORK_CIDR --name private \
  --dns-nameserver DNS_RESOLVER --gateway PRIVATE_NETWORK_GATEWAY
```

替换 PRIVATE_NETWORK_CIDR 为创建的子网 IP 地址范围。该 IP 地址值和范围可以是任意值，但是建议遵从 RFC 1918。

> **注意**：
> RFC 1918：私有地址网络分配，该分配允许一个企业内的所有主机之间及不同企业内的所有公开主机之间在网络所有层次上的连接。IANA（Internet Assigned Numbers Authority）建议使用以下地址区间作为私网地址：
> ```
>         10.0.0.0       ~    10.255.255.255   (10.0.0.0/8)
>         172.16.0.0     ~    172.31.255.255   (172.16.0.0/12)
>         192.168.0.0    ~    192.168.255.255  (192.168.0.0/16)
> ```

替换 DNS_RESOLVER 为 DNS 解析服务器的地址，一般该地址可参考计算节点的 /etc/resolv.conf 中的内容条目。

替换 PRIVATE_NETWORK_GATEWAY 为网关 IP 地址。一般该地址是以.1 结束的 IP 地址。

示例：公共网络的 IP 地址范围是 172.16.1.0/24，网关是 172.16.1.1，DHCP 服务器在 172.16.1.2～172.16.1.254 地址段内给虚拟机实例分配 IP 地址，DNS 解析地址是 8.8.4.4。

```
# neutron subnet-create private 172.16.1.0/24 --name private \
  --dns-nameserver 8.8.4.4 --gateway 172.16.1.1
```

Created a new subnet:

```
+-------------------+------------------------------------------------+
| Field             | Value                                          |
+-------------------+------------------------------------------------+
| allocation_pools  | {"start": "172.16.1.2", "end": "172.16.1.254"} |
| cidr              | 172.16.1.0/24                                  |
| dns_nameservers   | 8.8.4.4                                        |
| enable_dhcp       | True                                           |
| gateway_ip        | 172.16.1.1                                     |
| host_routes       |                                                |
| id                | 3482f524-8bff-4871-80d4-5774c2730728           |
| ip_version        | 4                                              |
| ipv6_address_mode |                                                |
| ipv6_ra_mode      |                                                |
| name              | private                                        |
| network_id        | 7c6f9b37-76b4-463e-98d8-27e5686ed083           |
| subnetpool_id     |                                                |
| tenant_id         | f5b2ccaa75ac413591f12fcaa096aa5c               |
+-------------------+------------------------------------------------+
```

（4）创建路由。

Private Project Network 通过虚拟路由连接到 Public Provider Network，每个路由都必须包含至少一个 Private Project Network 接口和 Public Provider Network 的网关。Public Provider Network 必须包含 router: external 使虚拟路由能够连接到外部网络，admin 用户或者其他具有管理员权限的用户在创建网络时必须包含此选项，当然，它也支持后期更新。

① 在控制节点上加载 admin user 的环境变量。

# source admin-openrc.sh

② 更新已经存在的 Public 网络配置信息。

# neutron net-update public --router:external
Updated network: public

③ 加载 demo user 的环境变量。

# source demo-openrc.sh

④ 创建路由。

# neutron router-create router
Created a new router:

```
+---------------------+------------------------------------------+
| Field               | Value                                    |
+---------------------+------------------------------------------+
| admin_state_up      | True                                     |
| external_gateway_info |                                        |
| id                  | 89dd2083-a160-4d75-ab3a-14239f01ea0b     |
| name                | router                                   |
| routes              |                                          |
| status              | ACTIVE                                   |
| tenant_id           | f5b2ccaa75ac413591f12fcaa096aa5c         |
+---------------------+------------------------------------------+
```

⑤ 关联子网和路由。

```
# neutron router-interface-add router private
Added interface bff6605d-824c-41f9-b744-21d128fc86e1 to router router.
```

⑥ 设置虚拟路由上 Public Provider Network 的网关。

```
# neutron router-gateway-set router public
Set gateway for router router
```

(5) 验证。

在进行下一步操作之前，需要对之前的操作进行检查，以修复错误。

① 在控制节点上加载 admin user 的环境变量。

```
# source admin-openrc.sh
```

② 列出网络 namespace。

```
# ip netns
qrouter-89dd2083-a160-4d75-ab3a-14239f01ea0b
qdhcp-7c6f9b37-76b4-463e-98d8-27e5686ed083
qdhcp-0e62efcd-8cee-46c7-b163-d8df05c3c5ad
```

③ 列出虚拟路由上的端口信息。

```
# neutron router-port-list router
```

```
+--------------------------------------+-------------------+----------------------------------------+
| id                                   | mac_address       | fixed_ips                              |
+--------------------------------------+-------------------+----------------------------------------+
| bff6605d-824c-41f9-b744-21d128fc86e1 | fa:16:3e:2f:34:9b | {"subnet_id":                          |
|                                      |                   | "3482f524-8bff-4871-80d4-5774c2730728",|
|                                      |                   | "ip_address": "172.16.1.1"}            |
| d6fe98db-ae01-42b0-a860-37b1661f5950 | fa:16:3e:e8:c1:41 | {"subnet_id":                          |
|                                      |                   | "5cc70da8-4ee7-4565-be53-b9c011fca011",|
|                                      |                   | "ip_address": "203.0.113.102"}         |
+--------------------------------------+-------------------+----------------------------------------+
```

④ 在配置有 Public Provider Network 的节点上 ping 该 IP 地址以检查正确性。

```
# ping -c 4 203.0.113.102
PING 203.0.113.102 (203.0.113.102) 56(84) bytes of data.
64 bytes from 203.0.113.102: icmp_req=1 ttl=64 time=0.619 ms
64 bytes from 203.0.113.102: icmp_req=2 ttl=64 time=0.189 ms
64 bytes from 203.0.113.102: icmp_req=3 ttl=64 time=0.165 ms
64 bytes from 203.0.113.102: icmp_req=4 ttl=64 time=0.216 ms

--- 203.0.113.102 ping statistics ---
rtt min/avg/max/mdev = 0.165/0.297/0.619/0.187 ms
```

## 20.2 创建 Key Pair

大部分 OpenStack 云镜像都支持公钥认证而不是传统的密码认证，因此，在创建虚拟机实例之前，必须为 Nova 计算服务添加一个公钥。

（1）加载 demo 租户环境变量。

```
# source demo-openrc.sh
```

（2）生成并添加 Key Pair。

```
# ssh-keygen -q -N ""
# nova keypair-add --pub-key ~/.ssh/id_rsa.pub mykey
```

（3）检查 Key Pair。

```
# nova keypair-list
```

```
+-------+-------------------------------------------------+
| Name  | Fingerprint                                     |
+-------+-------------------------------------------------+
| mykey | cc:e2:da:91:26:0f:85:81:bc:52:09:01:6e:ad:56:ad |
+-------+-------------------------------------------------+
```

## 20.3 创建 Security Group 规则

在 OpenStack 环境中包含一个默认的 Security Group，默认所有的虚拟机实例都可以使用它，但是默认的 Security Group 规则是拒绝一切访问。针对目前使用的 CirrOS Image，只开通 ping 和 SSH 两项功能。

(1) 对默认的 Security Group 添加 ping 规则。

```
# nova secgroup-add-rule default icmp -1 -1 0.0.0.0/0
+-------------+-----------+---------+-----------+--------------+
| IP Protocol | From Port | To Port | IP Range  | Source Group |
+-------------+-----------+---------+-----------+--------------+
| icmp        | -1        | -1      | 0.0.0.0/0 |              |
+-------------+-----------+---------+-----------+--------------+
```

(2) 对默认的 Security Group 添加 SSH 访问。

```
# nova secgroup-add-rule default tcp 22 22 0.0.0.0/0
+-------------+-----------+---------+-----------+--------------+
| IP Protocol | From Port | To Port | IP Range  | Source Group |
+-------------+-----------+---------+-----------+--------------+
| tcp         | 22        | 22      | 0.0.0.0/0 |              |
+-------------+-----------+---------+-----------+--------------+
```

## 20.4 创建虚拟机实例

根据已经选择的网络架构一或架构二创建虚拟机实例。

### 20.4.1 创建虚拟机实例（Public Provider Network）

在创建虚拟机实例之前，首先需要确定 Flavor（主机类型）、Image Name（镜像名称）、Network（网络）、Security Group（安全组）、Key Pair 和 Instance Name（虚拟机实例名称）。

(1) 在控制节点上加载 demo user 的环境变量。

```
# source demo-openrc.sh
```

（2）列出全部有效的 Flavor。

```
# nova flavor-list
+----+-----------+-----------+------+-----------+------+-------+-------------+-----------+
| ID | Name      | Memory_MB | Disk | Ephemeral | Swap | VCPUs | RXTX_Factor | Is_Public |
+----+-----------+-----------+------+-----------+------+-------+-------------+-----------+
| 1  | m1.tiny   | 512       | 1    | 0         |      | 1     | 1.0         | True      |
| 2  | m1.small  | 2048      | 20   | 0         |      | 1     | 1.0         | True      |
| 3  | m1.medium | 4096      | 40   | 0         |      | 2     | 1.0         | True      |
| 4  | m1.large  | 8192      | 80   | 0         |      | 4     | 1.0         | True      |
| 5  | m1.xlarge | 16384     | 160  | 0         |      | 8     | 1.0         | True      |
+----+-----------+-----------+------+-----------+------+-------+-------------+-----------+
```

（3）列出全部有效的 Image。

```
# nova image-list
+--------------------------------------+--------+--------+--------+
| ID                                   | Name   | Status | Server |
+--------------------------------------+--------+--------+--------+
| 0ee987b9-4a8e-4908-8ab5-833b781f0502 | cirros | ACTIVE |        |
+--------------------------------------+--------+--------+--------+
```

（4）列出全部有效的 Network。

```
# neutron net-list
+--------------------------------------+--------+-----------------------------------------------------+
| id                                   | name   | subnets                                             |
+--------------------------------------+--------+-----------------------------------------------------+
| 37193c68-b70c-4153-b5cf-b6f19e9d8951 | public | b51b9bdb-05d8-40c5-aa73-c70557bc46a9 10.96.1.0/24   |
|                                      |        | a60f0452-ba3d-41bd-83d7-86a38c12f5b5 1.1.11.0/24    |
+--------------------------------------+--------+-----------------------------------------------------+
```

> **注意：**
> 虚拟机实例使用 Public Provider Network，需要指定 Network ID，不能使用 Network 名称。

（5）列出全部有效的 Security Group。

```
# nova secgroup-list
+--------------------------------------+---------+------------------------+
| Id                                   | Name    | Description            |
+--------------------------------------+---------+------------------------+
| e4ad703c-f8de-44cd-bcc6-d111da6756fe | default | Default security group |
+--------------------------------------+---------+------------------------+
```

（6）创建实例。

```
# nova boot --flavor m1.tiny --image cirros --nic net-id=37193c68-b70c-4153-b5cf-b6f19e9d8951 \
    --security-group default --key-name mykey instance01
```

```
+---------------------------------------+------------------------------------------------------+
| Property                              | Value                                                |
+---------------------------------------+------------------------------------------------------+
| OS-DCF:diskConfig                     | MANUAL                                               |
| OS-EXT-AZ:availability_zone           | nova                                                 |
| OS-EXT-STS:power_state                | 1                                                    |
| OS-EXT-STS:task_state                 | -                                                    |
| OS-EXT-STS:vm_state                   | active                                               |
| OS-SRV-USG:launched_at                | 2016-03-14T03:35:41.000000                           |
| OS-SRV-USG:terminated_at              | -                                                    |
| accessIPv4                            |                                                      |
| accessIPv6                            |                                                      |
| config_drive                          |                                                      |
| created                               | 2016-03-14T03:35:18Z                                 |
| flavor                                | m1.tiny (1)                                          |
| hostId                                | a8ade516ce11f5cf4732fc51a2f8e655541ab9f8c3cf7229972a75ea |
| id                                    | 09bce5e1-0cad-4b19-a678-9cfe151237ff                 |
| image                                 | cirros (0ee987b9-4a8e-4908-8ab5-833b781f0502)        |
| key_name                              | mykey                                                |
| metadata                              | {}                                                   |
| name                                  | instance01                                           |
| os-extended-volumes:volumes_attached  | [{"id": "81e14efa-cc7c-4837-8763-d450a63a93ea"}]     |
| progress                              | 0                                                    |
| public network                        | 10.96.1.103                                          |
| security_groups                       | default                                              |
| status                                | ACTIVE                                               |
| tenant_id                             | 9543b7db999641fda39ff53dc22e7bcb                     |
| updated                               | 2016-03-14T03:35:42Z                                 |
| user_id                               | 33cd0cb0e80a4ba6bb5b8c36947f6267                     |
+---------------------------------------+------------------------------------------------------+
```

> **错误:**
>
> 执行此步骤后，查看创建虚拟机实例的 Console Log，发现错误。
>
> ```
> http://169.254.169.254/2009-04-04/instance-id
> failed 1/20: up 208.81. request failed
> failed 2/20: up 211.62. request failed
> failed 3/20: up 214.12. request failed
> failed 4/20: up 216.62. request failed
> failed 5/20: up 218.95. request failed
> failed 6/20: up 221.29. request failed
> ```
>
> 接着检查 Neutron 的 Log（neutron-dhcp-agent.log），报错如下。
>
> ```
> 2016-03-13  16:15:08.292  46929  INFO  neutron.agent.dhcp.agent  [-] Synchronizing state
> 2016-03-13 16:15:10.919 46929 ERROR neutron.agent.linux.utils [-]
> Command: ['sudo','neutron-rootwrap','/etc/neutron/rootwrap.conf', 'ip', 'netns', 'exec', 'qdhcp-37193c68-b70c-4153-b5cf-b6f19e9d8951', 'dnsmasq', '--no-hosts', '--no-resolv', '--strict-order', '--except-interface=lo', '--pid-file=/var/lib/neutron/dhcp/37193c68-b70c-4153-b5cf-b6f19e9d8951/pid','--dhcp-hostsfile=/var/lib/neutron/dhcp/37193c68-b70c-4153-b5cf-b6f19e9d8951/host',  '--addn-hosts=/var/lib/neutron/dhcp/37193c68-b70c-4153-b5cf-b6f19e9d8951/addn_hosts',  '--dhcp-optsfile=/var/lib/neutron/dhcp/37193c68-b70c-4153-b5cf-b6f19e9d8951/opts', '--dhcp-leasefile=/var/lib/neutron/dhcp/37193c68-b70c-4153-b5cf-b6f19e9d8951
> ```

```
/leases','--dhcp-match=set:ipxe,175', '--bind-interfaces', '--interface=
ns-6a946805-3e', '--dhcp-range=set:tag0,10.96.1.0,static,86400s', '--dhcp-
lease-max=256', '--conf-file=', '--domain=openstacklocal']
    Exit code: 3
    Stdin:
    Stdout:
    Stderr:
    dnsmasq: cannot open or create lease file /var/lib/neutron/dhcp/37193c68-
b70c-4153-b5cf-b6f19e9d8951/leases: Permission denied
    2016-03-13 16:15:10.928 46929 ERROR neutron.agent.dhcp.agent [-] Unable
to enable dhcp for 37193c68-b70c-4153-b5cf-b6f19e9d8951.
    2016-03-13 16:15:10.928 46929 ERROR neutron.agent.dhcp.agent Traceback
(most recent call last):
    2016-03-13 16:15:10.928 46929 ERROR neutron.agent.dhcp.agent File
"/usr/lib/python2.7/site-packages/neutron/agent/dhcp/agent.py", line 115,
in call_driver
    2016-03-13 16:15:10.928 46929 ERROR neutron.agent.dhcp.agent getattr
(driver, action)(**action_kwargs)
    2016-03-13 16:15:10.928 46929 ERROR neutron.agent.dhcp.agent File
"/usr/lib/python2.7/site-packages/neutron/agent/linux/dhcp.py", line 207,
in enable
    2016-03-13 16:15:10.928 46929 ERROR neutron.agent.dhcp.agent self.
spawn_process()
    2016-03-13 16:15:10.928 46929 ERROR neutron.agent.dhcp.agent File
"/usr/lib/python2.7/site-packages/neutron/agent/linux/dhcp.py", line 416,
in spawn_process
    2016-03-13 16:15:10.928 46929 ERROR neutron.agent.dhcp.agent self._
spawn_or_reload_process(reload_with_HUP=False)2016-03-13 16:15:10.928 46929
ERROR neutron.agent.dhcp.agent   File "/usr/lib/python2.7/site- packages/
neutron/agent/linux/dhcp.py", line 430, in _spawn_or_reload_process
    2016-03-13 16:15:10.928 46929 ERROR neutron.agent.dhcp.agent
pm.enable(reload_cfg=reload_with_HUP)
    2016-03-13 16:15:10.928 46929 ERROR neutron.agent.dhcp.agent File
"/usr/lib/python2.7/site-packages/neutron/agent/linux/external_process.p
y", line 92, in enable
    2016-03-13 16:15:10.928 46929 ERROR neutron.agent.dhcp.agent run_as_
root=self.run_as_root)
```

```
2016-03-13 16:15:10.928 46929 ERROR neutron.agent.dhcp.agent File"/usr/
lib/python2.7/site-packages/neutron/agent/linux/ip_lib.py", line 861, in
execute
   2016-03-13 16:15:10.928 46929 ERROR neutron.agent.dhcp.agent log_fail_
as_error=log_fail_as_error, **kwargs)
   2016-03-13 16:15:10.928 46929 ERROR neutron.agent.dhcp.agent File "/usr/
lib/python2.7/site-packages/neutron/agent/linux/utils.py", line 159, in
execute
   2016-03-13 16:15:10.928 46929 ERROR neutron.agent.dhcp.agent raise
RuntimeError(m)
   2016-03-13 16:15:10.928 46929 ERROR neutron.agent.dhcp.agent RuntimeError:
   2016-03-13 16:15:10.928 46929 ERROR neutron.agent.dhcp.agent Command:
['sudo', 'neutron-rootwrap', '/etc/neutron/rootwrap.conf', 'ip', 'netns',
'exec', 'qdhcp-37193c68-b70c-4153-b5cf-b6f19e9d8951', 'dnsmasq', '--no-
hosts', '--no-resolv', '--strict-order', '--except-interface=lo', '--pid-
   file=/var/lib/neutron/dhcp/37193c68-b70c-4153-b5cf-b6f19e9d8951/pid',
'--dhcp-hostsfile=/var/lib/neutron/dhcp/37193c68-b70c-4153-b5cf-b6f19e9d
8951/host', '--addn-hosts=/var/lib/neutron/dhcp/37193c68-b70c-4153-b5cf-
b6f19e9d8951/addn_hosts', '--dhcp-optsfile=/var/lib/neutron/dhcp/37193c68-
b70c-4153-b5cf-b6f19e9d8951/opts', '--dhcp-leasefile=/var/lib/neutron/
dhcp/37193c68-b70c-4153-b5cf-b6f19e9d8951/leases', '--dhcp-match=set:ipxe,
175', '--bind-interfaces', '--interface=ns-6a946805-3e', '--dhcp-range=set:
tag0,10.96.1.0,static,86400s', '--dhcp-lease-max=256', '--conf-file=',
'--domain=openstacklocal']
   2016-03-13 16:15:10.928 46929 ERROR neutron.agent.dhcp.agent Exit code: 3
   2016-03-13 16:15:10.928 46929 ERROR neutron.agent.dhcp.agent Stdin:
   2016-03-13 16:15:10.928 46929 ERROR neutron.agent.dhcp.agent Stdout:
   2016-03-13 16:15:10.928 46929 ERROR neutron.agent.dhcp.agent Stderr:
   **2016-03-13 16:15:10.928 46929 ERROR neutron.agent.dhcp.agent dnsmasq:
cannot open or create lease file /var/lib/neutron/dhcp/37193c68-b70c- 4153-
b5cf-b6f19e9d8951/leases: Permission denied**
   2016-03-13 16:15:10.928 46929 ERROR neutron.agent.dhcp.agent
   2016-03-13 16:15:10.928 46929 ERROR neutron.agent.dhcp.agent
   2016-03-13 16:15:10.932 46929 INFO neutron.agent.dhcp.agent [-]
Synchronizing state complete
```

此时已经明确地看到报错信息，提示访问 leases 文件时权限受限，原因可能有两个。

① 文件权限或者属主不正确。

```
-rw-r--r-- 1 neutron neutron 158 May  4 11:38
/var/lib/neutron/dhcp/37193c68-b70c-4153-b5cf-b6f19e9d8951/leases
```

文件权限和属主正确，排除该原因。

② 安全设置不正确。

- 关闭防火墙，已关闭。
- 关闭 SELinux，已关闭。
- 检查 Apparmor，输入 apparmor_status 命令，其中包含/usr/sbin/dnsmasq，原因基本确认，等待验证。

找到/etc/apparmor.d/usr.sbin.dnsmasq 文件，进行备份，然后删除该文件，重启服务器，生效。经测试，功能恢复正常。

(7) 检查实例状态。

```
# nova list
+--------------------------------------+------------+--------+------------+-------------+-------------------+
| ID                                   | Name       | Status | Task State | Power State | Networks          |
+--------------------------------------+------------+--------+------------+-------------+-------------------+
| 09bce5e1-0cad-4b19-a678-9cfe151237ff | instance01 | ACTIVE | -          | Running     | public=10.96.1.103|
+--------------------------------------+------------+--------+------------+-------------+-------------------+
```

**注意：**

虚拟机实例状态（Status 列）由 BUILD [第（6）步] 改为 ACTIVE [第（7）步]，证明已经成功创建虚拟机实例。

(8) 使用 Web 浏览器通过 VNC 访问实例。

```
# nova get-vnc-console instance01 novnc
+-------+-----------------------------------------------------------------------------------+
| Type  | Url                                                                               |
+-------+-----------------------------------------------------------------------------------+
| novnc | http://Controller:6080/vnc_auto.html?token=de8dc9c7-d60c-4d85-be19-272cdf0ffb06   |
+-------+-----------------------------------------------------------------------------------+
```

**注意：**

如果运行该 URL 的主机不能解析 Controller 名字，则替换成 IP 地址使用。

(9) 检查实例的网络连接。

```
# ping -c 4 10.96.1.103
PING 10.96.1.103 (10.96.1.103): 56 data bytes
64 bytes from 10.96.1.103: seq=0 ttl=64 time=8.365 ms
64 bytes from 10.96.1.103: seq=1 ttl=64 time=2.056 ms
64 bytes from 10.96.1.103: seq=2 ttl=64 time=2.590 ms
64 bytes from 10.96.1.103: seq=3 ttl=64 time=1.143 ms

--- 10.96.1.103 ping statistics ---
4 packets transmitted, 4 packets received, 0% packet loss
round-trip min/avg/max = 1.143/3.538/8.365 ms
```

(10) 通过 SSH 访问实例。

```
# ssh cirros@10.96.1.103

Host '10.96.1.103' is not in the trusted hosts file.
(fingerprint md5 5d:81:d6:3b:00:47:9f:80:bc:26:0a:9f:b6:81:f1:00)
Do you want to continue connecting? (y/n) yes
cirros@10.96.1.103's password:
$
```

注意:
CirrOS Image 使用的默认用户名和密码是 cirros 和 cubswin:)。

## 20.4.2 创建虚拟机实例（Private Project Network）

在创建虚拟机实例之前，首先需要确定 Flavor（主机类型）、Image Name（镜像名称）、Network（网络）、Security Group（安全组）、Key Pair 和 Instance Name（虚拟机实例名称）。

(1) 在控制节点上加载 demo user 的环境变量。

```
# source demo-openrc.sh
```

(2) 列出全部有效的 Flavor。

```
# nova flavor-list
```

```
+----+----------+-----------+------+-----------+------+-------+-------------+-----------+
| ID | Name     | Memory_MB | Disk | Ephemeral | Swap | VCPUs | RXTX_Factor | Is_Public |
+----+----------+-----------+------+-----------+------+-------+-------------+-----------+
| 1  | m1.tiny  | 512       | 1    | 0         |      | 1     | 1.0         | True      |
| 2  | m1.small | 2048      | 20   | 0         |      | 1     | 1.0         | True      |
| 3  | m1.medium| 4096      | 40   | 0         |      | 2     | 1.0         | True      |
| 4  | m1.large | 8192      | 80   | 0         |      | 4     | 1.0         | True      |
| 5  | m1.xlarge| 16384     | 160  | 0         |      | 8     | 1.0         | True      |
+----+----------+-----------+------+-----------+------+-------+-------------+-----------+
```

（3）列出全部有效的 Image。

```
# nova image-list
+--------------------------------------+--------+--------+--------+
| ID                                   | Name   | Status | Server |
+--------------------------------------+--------+--------+--------+
| 0ee987b9-4a8e-4908-8ab5-833b781f0502 | cirros | ACTIVE |        |
+--------------------------------------+--------+--------+--------+
```

（4）列出全部有效的 Network。

```
# neutron net-list
+--------------------------------------+---------+----------------------------------------------------+
| id                                   | name    | subnets                                            |
+--------------------------------------+---------+----------------------------------------------------+
| 0e62efcd-8cee-46c7-b163-d8df05c3c5ad | public  | 5cc70da8-4ee7-4565-be53-b9c011fca011 10.3.31.0/24  |
| 7c6f9b37-76b4-463e-98d8-27e5686ed083 | private | 3482f524-8bff-4871-80d4-5774c2730728 172.16.1.0/24 |
+--------------------------------------+---------+----------------------------------------------------+
```

> **注意：**
> 虚拟机实例使用 Private Project Network，需要指定 Network ID，不能使用 Network 名称。

（5）列出全部有效的 Security Group。

```
# nova secgroup-list
+--------------------------------------+---------+------------------------+
| Id                                   | Name    | Description            |
+--------------------------------------+---------+------------------------+
| e4ad703c-f8de-44cd-bcc6-d111da6756fe | default | Default security group |
+--------------------------------------+---------+------------------------+
```

（6）创建实例。

使用刚刚创建的 Network ID 替换 PRIVATE_NET_ID 字段。

```
# nova boot --flavor m1.tiny --image cirros --nic net-id=PRIVATE_NET_ID \
  --security-group default --key-name mykey private-instance
```

```
+-------------------------------------+-------------------------------------------------+
| Property                            | Value                                           |
+-------------------------------------+-------------------------------------------------+
| OS-DCF:diskConfig                   | MANUAL                                          |
| OS-EXT-AZ:availability_zone         | nova                                            |
| OS-EXT-STS:power_state              | 0                                               |
| OS-EXT-STS:task_state               | scheduling                                      |
| OS-EXT-STS:vm_state                 | building                                        |
| OS-SRV-USG:launched_at              | -                                               |
| OS-SRV-USG:terminated_at            | -                                               |
| accessIPv4                          |                                                 |
| accessIPv6                          |                                                 |
| adminPass                           | oMeLMk9zVGpk                                    |
| config_drive                        |                                                 |
| created                             | 2015-09-17T22:36:05Z                            |
| flavor                              | m1.tiny (1)                                     |
| hostId                              |                                                 |
| id                                  | 113c5892-e58e-4093-88c7-e33f502eaaa4            |
| image                               | cirros (38047887-61a7-41ea-9b49-27987d5e8bb9)   |
| key_name                            | mykey                                           |
| metadata                            | {}                                              |
| name                                | private-instance                                |
| os-extended-volumes:volumes_attached| []                                              |
| progress                            | 0                                               |
| security_groups                     | default                                         |
| status                              | BUILD                                           |
| tenant_id                           | f5b2ccaa75ac413591f12fcaa096aa5c                |
| updated                             | 2015-09-17T22:36:05Z                            |
| user_id                             | 684286a9079845359882afc3aa5011fb                |
+-------------------------------------+-------------------------------------------------+
```

（7）检查实例状态。

```
# nova list
+--------------------------------------+------------------+--------+------------+-------------+------------------------+
| ID                                   | Name             | Status | Task State | Power State | Networks               |
+--------------------------------------+------------------+--------+------------+-------------+------------------------+
| 113c5892-e58e-4093-88c7-e33f502eaaa4 | private-instance | ACTIVE | -          | Running     | private=172.16.1.3     |
| 181c52ba-aebc-4c82-a97d-2e8e82e4eaaf | public-instance  | ACTIVE | -          | Running     | public=203.0.113.103   |
+--------------------------------------+------------------+--------+------------+-------------+------------------------+
```

> 注意：
> 虚拟机实例状态（Status 列）由 BUILD［第（6）步］改为 ACTIVE［第（7）步］，证明已经成功创建虚拟机实例。

（8）使用 Web 浏览器通过 VNC 访问实例。

```
# nova get-vnc-console private-instance novnc
```

```
+--------+---------------------------------------------------------------------------+
| Type   | Url                                                                       |
+--------+---------------------------------------------------------------------------+
| novnc  | http://controller:6080/vnc_auto.html?token=2f6dd985-f906-4bfc-b566-e87ce656375b |
+--------+---------------------------------------------------------------------------+
```

**注意：**
如果运行该 URL 的主机不能解析 Controller 名字，则替换成 IP 地址使用。

（9）检查实例 Private Project Network 网关连接。

```
# ping -c 4 172.16.1.1
PING 172.16.1.1 (172.16.1.1) 56(84) bytes of data.
64 bytes from 172.16.1.1: icmp_req=1 ttl=64 time=0.357 ms
64 bytes from 172.16.1.1: icmp_req=2 ttl=64 time=0.473 ms
64 bytes from 172.16.1.1: icmp_req=3 ttl=64 time=0.504 ms
64 bytes from 172.16.1.1: icmp_req=4 ttl=64 time=0.470 ms

--- 172.16.1.1 ping statistics ---
4 packets transmitted, 4 received, 0% packet loss, time 2998ms
rtt min/avg/max/mdev = 0.357/0.451/0.504/0.055 ms
```

（10）检查实例是否可以访问互联网。

```
# ping -c 4 openstack.org
PING openstack.org (174.143.194.225) 56(84) bytes of data.
64 bytes from 174.143.194.225: icmp_req=1 ttl=53 time=17.4 ms
64 bytes from 174.143.194.225: icmp_req=2 ttl=53 time=17.5 ms
64 bytes from 174.143.194.225: icmp_req=3 ttl=53 time=17.7 ms
64 bytes from 174.143.194.225: icmp_req=4 ttl=53 time=17.5 ms

--- openstack.org ping statistics ---
4 packets transmitted, 4 received, 0% packet loss, time 3003ms
rtt min/avg/max/mdev = 17.431/17.575/17.734/0.143 ms
```

（11）在 Public Provider Network 创建 Floating IP 地址。

```
# neutron floatingip-create public
Created a new floatingip:
```

```
+------------------------+--------------------------------------+
| Field                  | Value                                |
+------------------------+--------------------------------------+
| fixed_ip_address       |                                      |
| floating_ip_address    | 203.0.113.104                        |
| floating_network_id    | 9bce64a3-a963-4c05-bfcd-161f708042d1 |
| id                     | 05e36754-e7f3-46bb-9eaa-3521623b3722 |
| port_id                |                                      |
| router_id              |                                      |
| status                 | DOWN                                 |
| tenant_id              | 7cf50047f8df4824bc76c2fdf66d11ec     |
+------------------------+--------------------------------------+
```

（12）分配 Floating IP 地址给虚拟机实例。

```
# nova floating-ip-associate private-instance 203.0.113.104
```

**注意：**

此命令执行完成后，没有输出。

（13）检查 Floating IP 地址状态。

```
# nova list
```
```
+--------------------------------------+------------------+--------+------------+-------------+----------------------------------------+
| ID                                   | Name             | Status | Task State | Power State | Networks                               |
+--------------------------------------+------------------+--------+------------+-------------+----------------------------------------+
| 113c5892-e58e-4098-88c7-e831502eaaa4 | private-instance | ACTIVE | -          | Running     | private=172.16.1.3, 203.0.113.104      |
| 181c52ba-aebc-4c82-a97d-2e8e82e4eaaf | public-instance  | ACTIVE | -          | Running     | public=203.0.113.103                   |
+--------------------------------------+------------------+--------+------------+-------------+----------------------------------------+
```

（14）在配置 Public Provider Network 的服务器上检查 Floating IP 地址的连接情况。

```
# ping -c 4 203.0.113.104
PING 203.0.113.104 (203.0.113.104) 56(84) bytes of data.
64 bytes from 203.0.113.104: icmp_req=1 ttl=63 time=3.18 ms
64 bytes from 203.0.113.104: icmp_req=2 ttl=63 time=0.981 ms
64 bytes from 203.0.113.104: icmp_req=3 ttl=63 time=1.06 ms
64 bytes from 203.0.113.104: icmp_req=4 ttl=63 time=0.929 ms

--- 203.0.113.104 ping statistics ---
4 packets transmitted, 4 received, 0% packet loss, time 3002ms
rtt min/avg/max/mdev = 0.929/1.539/3.183/0.951 ms
```

（15）通过 SSH 访问实例。

```
# ssh cirros@203.0.113.104
```

```
Host '203.0.113.104' is not in the trusted hosts file.
(fingerprint md5 5d:8d:d7:3a:34:42:7y:89:bc:26:0a:9f:b6:64:h4:17)
Do you want to continue connecting? (y/n) yes
cirros@203.0.113.104's password:
$
```

> **注意：**
> CirrOS Image 使用的默认用户名和密码是 cirros 和 cubswin:)。

## 20.5 创建块存储

下面开始创建卷设备，供虚拟机实例使用。

（1）在控制节点上加载 demo user 的环境变量。

```
# source demo-openrc.sh
```

（2）创建 1GB 的卷设备。

```
# cinder create --display-name volume01 1
```

（3）检查卷设备的状态。

```
# cinder list
```

(4) 分配卷设备给虚拟机实例。

`# nova volume-attach INSTANCE_NAME VOLUME_ID`

替换 INSTANCE_NAME 和 VOLUME_ID 为虚拟机实例名字和卷设备 ID。

举例：分配 81e14efa-cc7c-4837-8763-d450a63a93ea volume 给 instance01。

`# nova volume-attach instance01 81e14efa-cc7c-4837-8763-d450a63a93ea`

```
+----------+----------------------------------------+
| Property | Value                                  |
+----------+----------------------------------------+
| device   | /dev/vdb                               |
| id       | 19ubea89-07db-4ac2-8115-66c0d6a4bb86   |
| serverId | 09bce5e1-0cad-4b19-a678-9cfe151237ff   |
| volumeId | 81e14efa-cc7c-4837-8763-d450a63a93ea   |
+----------+----------------------------------------+
```

> **错误：**
>
> 执行此步骤报错，检查 Cinder 的 Log 文件/var/log/cinder/volume.log，发现以下错误。
>
> ```
>     2016-03-14 15:51:22.907 12466 ERROR cinder.volume.targets.tgt [req-891b70dc-12e3-401e-bca5-d3ee9701052b 33cd0cb0e80a4ba6bb5b8c36947f6267 9543b7db999641fda39ff53dc22e7bcb - - -] Failed to create iscsi target for Volume ID: volume-81e14efa-cc7c-4837-8763-d450a63a93ea. Please ensure your tgtd config file contains 'include /var/lib/cinder/volumes/*'
>     2016-03-14 15:51:22.912 12466 ERROR cinder.volume.manager [req-891b70dc-12e3-401e-bca5-d3ee9701052b 33cd0cb0e80a4ba6bb5b8c36947f6267
>     9543b7db999641fda39ff53dc22e7bcb - - -] Create export for volume failed.
>     2016-03-14 15:51:22.912 12466 ERROR cinder.volume.manager Traceback (most recent call last):
>     2016-03-14 15:51:22.912 12466 ERROR cinder.volume.manager File "/usr/lib/python2.7/site-packages/cinder/volume/manager.py", line 1318, in initialize_connection
>     2016-03-14 15:51:22.912 12466 ERROR cinder.volume.manager volume, connector)
>     2016-03-14 15:51:22.912 12466 ERROR cinder.volume.manager File "/usr/lib/python2.7/site-packages/osprofiler/profiler.py", line 105, in wrapper
>     2016-03-14 15:51:22.912 12466 ERROR cinder.volume.manager return f(*args, **kwargs)
> ```

```
2016-03-14 15:51:22.912 12466 ERROR cinder.volume.manager File "/usr/lib/python2.7/site-packages/cinder/volume/drivers/lvm.py", line 725, in create_export
    2016-03-14 15:51:22.912 12466 ERROR cinder.volume.manager volume_path)
    2016-03-14 15:51:22.912 12466 ERROR cinder.volume.manager File "/usr/lib/python2.7/site-packages/cinder/volume/targets/iscsi.py", line 214, in create_export
    2016-03-14 15:51:22.912 12466 ERROR cinder.volume.manager **portals_config)
    2016-03-14 15:51:22.912 12466 ERROR cinder.volume.manager File "/usr/lib/python2.7/site-packages/cinder/volume/targets/tgt.py", line 257, in create_iscsi_target
    2016-03-14 15:51:22.912 12466 ERROR cinder.volume.manager raise exception.NotFound()
    2016-03-14 15:51:22.912 12466 ERROR cinder.volume.manager NotFound: Resource could not be found.
    2016-03-14 15:51:22.912 12466 ERROR cinder.volume.manager
    2016-03-14 15:51:22.922 12466 ERROR oslo_messaging.rpc.dispatcher [req-891b70dc-12e3-401e-bca5-d3ee9701052b 33cd0cb0e80a4ba6bb5b8c36947f62679543b7db999641fda39ff53dc22e7bcb - - -] Exception during message handling: Bad or unexpected response from the storage volume backend API: Create export for volume failed.
    2016-03-14 15:51:22.922 12466 ERROR oslo_messaging.rpc.dispatcher Traceback (most recent call last):
    2016-03-14 15:51:22.922 12466 ERROR oslo_messaging.rpc.dispatcher File "/usr/lib/python2.7/site-packages/oslo_messaging/rpc/dispatcher.py", line 142, in _dispatch_and_reply
    2016-03-14 15:51:22.922 12466 ERROR oslo_messaging.rpc.dispatcher executor_callback))
    2016-03-14 15:51:22.922 12466 ERROR oslo_messaging.rpc.dispatcher File "/usr/lib/python2.7/site-packages/oslo_messaging/rpc/dispatcher.py", line 186, in _dispatch
    2016-03-14 15:51:22.922 12466 ERROR oslo_messaging.rpc.dispatcher executor_callback)
    2016-03-14 15:51:22.922 12466 ERROR oslo_messaging.rpc.dispatcher File "/usr/lib/python2.7/site-packages/oslo_messaging/rpc/dispatcher.py", line 129, in _do_dispatch
```

```
    2016-03-14  15:51:22.922  12466  ERROR  oslo_messaging.rpc.dispatcher
result = func(ctxt, **new_args)
    2016-03-14  15:51:22.922  12466  ERROR  oslo_messaging.rpc.dispatcher
File "/usr/lib/python2.7/site-packages/osprofiler/profiler.py", line 105,
in wrapper
    2016-03-14  15:51:22.922  12466  ERROR  oslo_messaging.rpc.dispatcher
return f(*args, **kwargs)
    2016-03-14  15:51:22.922  12466  ERROR  oslo_messaging.rpc.dispatcher
File  "/usr/lib/python2.7/site-packages/cinder/volume/manager.py",  line
1322, in initialize_connection
    2016-03-14  15:51:22.922  12466  ERROR  oslo_messaging.rpc.dispatcher
raise exception.VolumeBackendAPIException(data=err_msg)
    2016-03-14  15:51:22.922  12466  ERROR  oslo_messaging.rpc.dispatcher
VolumeBackendAPIException: Bad or unexpected response from the storage volume
backend API: Create export for volume failed.
    2016-03-14 15:51:22.922 12466 ERROR oslo_messaging.rpc.dispatcher
    2016-03-14  15:51:22.930  12466  ERROR  oslo_messaging._drivers.common
[req-891b70dc-12e3-401e-bca5-d3ee9701052b
33cd0cb0e80a4ba6bb5b8c36947f6267
9543b7db999641fda39ff53dc22e7bcb - - -] Returning exception Bad or unexpected
response from the storage volume backend API: Create export for volume failed.
to caller
    2016-03-14  15:51:22.931  12466  ERROR  oslo_messaging._drivers.common
[req-891b70dc-12e3-401e-bca5-d3ee9701052b 33cd0cb0e80a4ba6bb5b8c 36947f6267
9543b7db999641fda39ff53dc22e7bcb  -  -  -]  ['Traceback  (most  recent  call
last):\n', ' File "/usr/lib/python2.7/site-packages/oslo_messaging/ rpc/
dispatcher.py", line 142, in _dispatch_and_reply\n executor_ callback))\n',
' File "/usr/lib/python2.7/site-packages/oslo_messaging/ rpc/dispatcher.
py", line 186, in _dispatch\n executor_callback)\n', ' File "/usr/lib/
python2.7/site-
packages/oslo_messaging/rpc/dispatcher.py", line 129, in _do_dispatch\n
result = func(ctxt, **new_args)\n', '  File "/usr/lib/python2.7/site-
packages/osprofiler/profiler.py", line 105, in wrapper\n    return f(*args,
**kwargs)\n',  '  File  "/usr/lib/python2.7/site-packages/cinder/volume/
manager.py", line 1322, in initialize_connection\n  raise exception.
VolumeBackendAPIException(data=err_msg)\n','VolumeBackendAPIException:
Bad or unexpected response from the storage volume backend API: Create export
```

```
for volume failed.\n']
```

错误已经很明显,需要向文件/etc/tgt/targets.conf 添加内容 "include /var/lib/cinder/volumes/*",然后重启 tgtd 和 openstack-cinder-volume 服务。经测试,功能恢复正常。

(5)列出 volume 的状态。

```
# nova volume-list
```

| ID | Status | Display Name | Size | Volume Type | Attached to |
|---|---|---|---|---|---|
| 81e14efa-cc7c-4837-8763-d450a63a93ea | in-use | volume01 | 1 | - | 09bce5e1-0cad-4b19-a678-9cfe151237ff |

(6)登录虚拟机实例操作系统,检查分配的 volume 状态是否正常。

```
$ sudo fdisk -l

Disk /dev/vda: 1073 MB, 1073741824 bytes
255 heads, 63 sectors/track, 130 cylinders, total 2097152 sectors
Units = sectors of 1 * 512 = 512 bytes
Sector size (logical/physical): 512 bytes / 512 bytes
I/O size (minimum/optimal): 512 bytes / 512 bytes
Disk identifier: 0x00000000

   Device Boot      Start         End      Blocks   Id  System
/dev/vda1   *       16065     2088449     1036192+  83  Linux

Disk /dev/vdb: 1073 MB, 1073741824 bytes
16 heads, 63 sectors/track, 2080 cylinders, total 2097152 sectors
Units = sectors of 1 * 512 = 512 bytes
Sector size (logical/physical): 512 bytes / 512 bytes
I/O size (minimum/optimal): 512 bytes / 512 bytes
Disk identifier: 0x00000000

Disk /dev/vdb doesn't contain a valid partition table
# df -h
Filesystem                Size      Used Available Use% Mounted on
/dev                    242.3M         0    242.3M   0% /dev
/dev/vda1                23.2M     18.0M      4.0M  82% /
tmpfs                   245.8M         0    245.8M   0% /dev/shm
tmpfs                   200.0K     72.0K    128.0K  36% /run
/dev/vdb                1007.9M    971.7M         0 100% /test
```

**注意：**

在使用卷设备（/dev/vdb）之前，需要创建文件系统并挂载。

举例：

/dev/vdb                  1007.9M      971.7M         0 100% /test
/dev/vdb on /test type ext3 (rw,relatime,errors=continue,user_xattr,acl,barrier=1,data=ordered)

# 第 3 篇

# 管 理 篇

- 第 21 章 OpenStack 项目管理
- 第 22 章 仪表板使用
- 第 23 章 管理镜像
- 第 24 章 管理网络
- 第 25 章 管理卷设备
- 第 26 章 管理虚拟机实例
- 第 27 章 OpenStack 版本升级
- 第 28 章 故障排查

# 第 21 章 OpenStack 项目管理

## 21.1 管理租户、用户和角色

租户（project）是 OpenStack 项目中的一个组织单元，用户（user）可以同时属于一个或多个租户，角色（role）定义了用户可以执行的操作类型。作为具有权限的管理员，可以管理所有的租户、用户和角色。

OpenStack 项目中的各个服务组件可以执行的操作可以在/etc/PROJECT/policy.json 文件中进行定义。例如，Nova 计算服务组件可执行操作列表在/etc/nova/policy.json 文件中定义。

租户、用户和角色彼此之间没有依赖性，可以分别进行操作。在建立 OpenStack 项目时，至少要创建一个租户、用户和角色。在进行删除用户的操作时，需要首先将用户和租户的映射删除。

注意：
对于一名技术人员来讲，具备清晰的逻辑思维很重要。例如，上述删除用户的操作，可以剖析其中的逻辑关系。
创建过程：
（1）建立租户。

（2）建立用户。

（3）建立角色。

（4）将用户赋予租户。

（5）将角色赋予该租户的该用户（一个用户可以同时属于多个租户）。

删除过程：

（1）删除用户与角色的映射关系。

（2）删除用户与租户的映射关系。

（3）删除角色。

（4）删除用户。

（5）删除租户。

这样操作就不会存在混乱的关系或僵尸事务。在传统的架构中，经常会遇到在操作系统上删除 SAN 存储磁盘的工作，也存在相似性。

识别 SAN 存储磁盘的过程：

（1）光纤交换机建立 Zone。

（2）存储创建 LUN。

（3）主机识别存储磁盘。

删除 SAN 存储磁盘的过程：

（1）主机删除存储磁盘。

（2）存储删除 LUN。

（3）光纤交换机删除 Zone。

这样做会避免操作系统上的磁盘混乱或鬼盘，减少出现问题的可能性，在很大程度上简化工作的复杂性、提高工作效率。

## 21.1.1 命令行方式

租户可以含有多个用户或者一个都没有，在 Nova 计算服务中，租户拥有多个虚拟机实例；在 Swift 对象存储服务中，租户拥有多个 Container。一个用户可以同时属于一个或者多个租户，因此，角色属于某个租户的某个用户，而不是单纯属于某个用户。

(1) 列表展示所有租户。

```
# openstack project list
+----------------------------------+---------+
| ID                               | Name    |
+----------------------------------+---------+
| 8af2565380aa40f1828423d892b60a7e | service |
| 9543b7db999641fda39ff53dc22e7bcb | demo    |
| efb33dccf25b40b399e92d6351fc7ea9 | admin   |
+----------------------------------+---------+
```

(2) 创建租户。

```
# openstack project create --description 'Demo Project' demo
+-------------+----------------------------------+
| Field       | Value                            |
+-------------+----------------------------------+
| description | Demo Project                     |
| domain_id   | default                          |
| enabled     | True                             |
| id          | 9543b7db999641fda39ff53dc22e7bcb |
| is_domain   | False                            |
| name        | demo                             |
| parent_id   | None                             |
+-------------+----------------------------------+
```

(3) 更新租户。

① 禁用租户。

```
# openstack project set PROJECT_ID --disable
```

② 启用被禁用的租户。

```
# openstack project set PROJECT_ID --enable
```

③ 更新租户名称。

```
# openstack project set PROJECT_ID --name project-new
```

④ 确认已修改的租户信息。

```
# openstack project show PROJECT_ID
+-------------+----------------------------------+
| Field       | Value                            |
+-------------+----------------------------------+
| description | Demo Project                     |
| domain_id   | default                          |
| enabled     | True                             |
| id          | 9543b7db999641fda39ff53dc22e7bcb |
| is_domain   | False                            |
| name        | project-new                      |
| parent_id   | None                             |
+-------------+----------------------------------+
```

(4) 删除租户。

```
# openstack project delete PROJECT_ID
```

(5) 列表展示所有用户。

```
# openstack user list
```

```
+----------------------------------+---------+
| ID                               | Name    |
+----------------------------------+---------+
| 09aa4d42b11648f29845d4b5e0120283 | neutron |
| 0c054bcf0bf54016966c5f096cc05add | cinder  |
| 11839df86ee94ca7a36ee514a4eaf316 | nova    |
| 2c27875abf014b33949cf5de9ebdf578 | glance  |
| 33cd0cb0e80a4ba6bb5b8c36947f6267 | demo    |
| fc3820e9803f46308092b409d6323392 | admin   |
+----------------------------------+---------+
```

(6) 创建用户。

在创建用户时，必须指定用户名称。可选项包括租户 ID 和密码，建议在创建用户时指定可选项信息，否则用户无法通过 dashboard 登录。

```
# openstack user create --domain default --password-prompt demo
User Password:
Repeat User Password:
```

```
+-----------+----------------------------------+
| Field     | Value                            |
+-----------+----------------------------------+
| domain_id | default                          |
| enabled   | True                             |
| id        | 33cd0cb0e80a4ba6bb5b8c36947f6267 |
| name      | demo                             |
+-----------+----------------------------------+
```

(7) 更新用户。

① 禁用用户。

```
# openstack user set USER_NAME --disable
```

② 启用被禁用的用户。

```
# openstack user set USER_NAME --enable
```

③ 更新用户名称。

```
# openstack user set USER_NAME --name user-new
```

(8) 删除用户。

```
# openstack user delete USER_NAME
```

(9) 列表展示所有角色。

```
# openstack role list
+----------------------------------+-------+
| ID                               | Name  |
+----------------------------------+-------+
| 361c58c26c8c46b49315fb33a3ad5870 | user  |
| e089ff227c504e07bda544bc07b0b989 | admin |
+----------------------------------+-------+
```

(10) 创建角色。

```
# openstack role create new-role
+-------+----------------------------------+
| Field | Value                            |
+-------+----------------------------------+
| id    | 361c58c26c8c46b49315fb33a3ad5870 |
| name  | user                             |
+-------+----------------------------------+
```

(11) 分配角色。

① 列出所有租户和用户并挑选所需赋予角色的用户。

```
# openstack project list
+----------------------------------+---------+
| ID                               | Name    |
+----------------------------------+---------+
| 8af2565380aa40f1828423d892b60a7e | service |
| 9543b7db999641fda39ff53dc22e7bcb | demo    |
| efb33dccf25b40b399e92d6351fc7ea9 | admin   |
+----------------------------------+---------+
```

```
# openstack user list
+----------------------------------+---------+
| ID                               | Name    |
+----------------------------------+---------+
| 09aa4d42b11648f29845d4b5e0120283 | neutron |
| 0c054bcf0bf54016966c5f096cc05add | cinder  |
| 11839df86ee94ca7a36ee514a4eaf316 | nova    |
| 2c27875abf014b33949cf5de9ebdf578 | glance  |
| 33cd0cb0e80a4ba6bb5b8c36947f6267 | demo    |
| fc3820e9803f46308092b409d6323392 | admin   |
+----------------------------------+---------+
```

② 列出所有角色并挑选所需使用的角色。

```
# openstack role list
```

```
+----------------------------------+-------+
| ID                               | Name  |
+----------------------------------+-------+
| 361c58c26c8c46b49315fb33a3ad5870 | user  |
| e089ff227c504e07bda544bc07b0b989 | admin |
+----------------------------------+-------+
```

③ 将角色赋予用户。

```
# openstack role add --user USER_NAME --project TENANT_ID ROLE_NAME
```

④ 确认结果。

```
# openstack role list --user demo --project demo
+----------------------------------+------+---------+------+
| ID                               | Name | Project | User |
+----------------------------------+------+---------+------+
| 361c58c26c8c46b49315fb33a3ad5870 | user | demo    | demo |
+----------------------------------+------+---------+------+
```

（12）删除角色。

```
# openstack role remove --user USER_NAME --project TENANT_ID ROLE_NAME
```

## 21.1.2 图形界面方式

角色是拥有权限的集合，定义用户可以执行的一系列操作。

### 1．创建角色

（1）登录 dashboard。

（2）在 Web 界面左侧，打开 Identity（身份管理）标签，选中 role（角色）分类，可以看到管理的所有角色信息。

（3）单击 Create Role（创建角色）按钮，在弹出的对话框中填写角色名称。

（4）在弹出的对话框中单击 Create Role（创建角色）按钮进行创建，如下图所示。

## 2. 编辑角色

（1）登录 dashboard。

（2）在 Web 界面左侧，打开 Identity（身份管理）标签，选中 role（角色）分类，可以看到管理的所有角色信息。

（3）单击 Edit（编辑）按钮，在弹出的对话框中填写角色名称。

（4）在弹出的 Update Role（更新角色）对话框中，输入角色的新名称。

（5）单击 Update Role（更新角色）按钮进行更新，如下图所示。

注意：

在 dashboard 中，只能更新角色的名称。

## 3. 删除角色

（1）登录 dashboard。

（2）在 Web 界面左侧，打开 Identity（身份管理）标签，选中 role（角色）分类，可以看到管理的所有角色信息。

（3）选中要删除的角色名称，单击 Delete Roles（删除角色）按钮。

（4）在弹出的对话框中单击 Confirm（确认）按钮进行删除。

警告：

此操作不能回退。

## 21.2 管理主机类型

主机类型定义了虚拟机实例运行时所需的硬件配置资源，包括 CPU、内存和存储容量大小。

OpenStack 项目中默认创建的主机类型如下表所示。

| 主机类型 | 虚拟内核 | 根磁盘(GB) | 内存(MB) |
|---|---|---|---|
| m1.tiny | 1 | 1 | 512 |
| m1.small | 1 | 20 | 2048 |
| m1.medium | 2 | 20 | 4096 |
| m1.large | 4 | 80 | 8192 |
| m1.xlarge | 8 | 160 | 16384 |

### 21.2.1 命令行方式

使用 nova flavor-* 命令管理主机类型。主机类型包括以下参数。

- 主机类型 ID：唯一的主机类型 ID，可以是整数，也可以是自动生成的 UUID。

- 主机类型名称：主机类型标识字段。

- 虚拟内核：需要使用的虚拟 CPU 核数。

- 内存：需要使用的内存大小（MB）。

- 根磁盘：虚拟机实例的根（/）文件系统所需的磁盘空间（GB）。

- 临时磁盘：虚拟机实例使用的临时磁盘空间，默认值为 0。临时磁盘来源于本地存储，供虚拟机实例使用。当关闭虚拟机实例时，临时磁盘中的数据会丢失。临时磁盘不能包括任何快照。

- 交换空间：虚拟机实例需要使用的交换空间（MB），默认值为 0。

## 1. 列表显示所有的主机类型

```
# nova flavor-list
+----+-----------+-----------+------+-----------+------+-------+-------------+-----------+
| ID | Name      | Memory_MB | Disk | Ephemeral | Swap | VCPUs | RXTX_Factor | Is_Public |
+----+-----------+-----------+------+-----------+------+-------+-------------+-----------+
| 1  | m1.tiny   | 512       | 1    | 0         |      | 1     | 1.0         | True      |
| 2  | m1.small  | 2048      | 20   | 0         |      | 1     | 1.0         | True      |
| 3  | m1.medium | 4096      | 40   | 0         |      | 2     | 1.0         | True      |
| 4  | m1.large  | 8192      | 80   | 0         |      | 4     | 1.0         | True      |
| 5  | m1.xlarge | 16384     | 160  | 0         |      | 8     | 1.0         | True      |
+----+-----------+-----------+------+-----------+------+-------+-------------+-----------+
```

## 2. 创建主机类型

（1）指定新建主机类型的名称、ID、内存、根磁盘和虚拟内核数量。

```
# nova flavor-create FLAVOR_NAME FLAVOR_ID RAM_IN_MB ROOT_DISK_IN_GB NUMBER_OF_VCPUS
```

 注意：

创建主机类型时指定的 ID 可以是手工定义的整数或者自动生成的 UUID。ID 参数为 auto 时，自动生成 UUID。

以下举例说明通过添加参数定义主机类型。主机名称为 m1.extra_tiny，自动生成 ID，内存为 256MB，没有磁盘，有一颗虚拟内核，rxtx-factor 参数定义虚拟网卡的带宽。

```
# nova flavor-create --is-public true m1.extra_tiny auto 256 0 1 -rxtx
-factor .1
```

（2）个人用户可以创建主机类型，分配给某租户，使其成为私有的主机类型。

```
# nova flavor-access-add FLAVOR TENANT_ID
```

（3）通过设置 extra_spec 参数可以为已有的主机类型设置更多的限制。例如，主机类型设置为 extra_spec key/value quota:vif_outbound_peak=65536，使用该主机类型的虚拟机实例传输 I/O 峰值带宽为 LTE 512 Mbit/s。还有一些参数可以影响虚拟机实例的使用，包括 CPU 限制、磁盘调优、I/O 带宽、Watchdog 操作和随机数生成器（Random-Number Generator）。具体信息可以通过以下方式查看。

```
# nova help flavor-key
```

### 3. 删除主机类型

```
# nova flavor-delete FLAVOR_ID
```

## 21.2.2 图形界面方式

在图形界面操作中，管理员可以创建、编辑和删除主机类型。

### 1. 创建主机类型

（1）登录 dashboard。

（2）在 Web 界面左侧，在 Admin（管理员）标签上，单击 System（系统）标签，选中 Flavor（云主机类型）分类，可以看到已定义的主机类型。

（3）单击 Create Flavor（创建云主机类型）按钮，输入或选择相关参数，如下图所示。

（4）在 Flavor Access（云主机类型访问）标签中，设置可以访问主机类型的租户列表，如下图所示。

[图示：创建云主机类型对话框]

将页面左面 All Project（所有项目）中的租户移动到右面 Selected Projects（选中的项目）部分，这样，只有右面的租户可以访问该主机类型。如果右面 Selected Projects（选中的项目）为空，则所有租户都可以访问该主机类型。

（5）单击 Create Flavor（创建云主机类型）按钮进行创建。

### 2. 更新主机类型

（1）登录 dashboard。

（2）在 Web 界面左侧，在 Admin（管理员）标签上，单击 System（系统）标签，选中 Flavor（云主机类型）分类，可以看到已定义的主机类型。

（3）选择需编辑的云主机类型，单击 Edit Flavor（编辑云主机类型）按钮。

（4）在弹出的对话框中，可以编辑云主机类型的名称、内存和磁盘等。

（5）单击 Save（保存）按钮，如下图所示。

## 3．更新 metadata

（1）登录 dashboard。

（2）在 Web 界面左侧，在 Admin（管理员）标签上，单击 System（系统）标签，选中 Flavor（云主机类型）分类，可以看到已定义的主机类型。

（3）选择需更新 metadata 的主机类型，在下拉列表中选择更新 metadata。

（4）在弹出的对话框中，可以填写一些 metadata 值。

（5）单击 Save（保存）按钮。

可选的 metadata 值如下表所示。

| | |
|---|---|
| CPU limits | quota:cpu_shares |
| | quota:cpu_period |
| | quota:cpu_limit |
| | quota:cpu_reservation |
| | quota:cpu_quota |

| | 续表 |
|---|---|
| Disk tuning | quota:disk_read_bytes_sec |
| | quota:disk_read_iops_sec |
| | quota:disk_write_bytes_sec |
| | quota:disk_write_iops_sec |
| | quota:disk_total_bytes_sec |
| | quota:disk_total_iops_sec |
| Bandwidth I/O | quota:vif_inbound_average |
| | quota:vif_inbound_burst |
| | quota:vif_inbound_peak |
| | quota:vif_outbound_average |
| | quota:vif_outbound_burst |
| | quota:vif_outbound_peak |
| Watchdog behavior | hw:watchdog_action |
| Random-number generator | hw_rng:allowed |
| | hw_rng:rate_bytes |
| | hw_rng:rate_period |

### 4. 删除主机类型

（1）登录 dashboard。

（2）在 Web 界面左侧，在 Admin（管理员）标签上，单击 System（系统）标签，选中 Flavor（云主机类型）分类，可以看到已定义的主机类型。

（3）选中需删除的主机类型。

（4）单击 Delete Flavor（删除云主机类型）按钮，在弹出的对话框中确认删除。

## 21.3 管理安全组

安全组是一系列 IP 过滤规则的集合，控制对虚拟机实例的网络访问规则，可以应用于租户内的所有虚拟机实例。管理员可以修改安全组内的规则。

所有的租户内都有一个默认的安全组，建立虚拟机实例时默认使用该安全组，当然也可以指定其他安全组。该安全组的默认规则是拒绝一切向虚拟机实例内的访问，允许从虚拟机实例向外的所有访问。

对于同一网络的主机节点,可以在/etc/nova/nova.conf 文件中定义 allow_same_net_traffic 参数以共享相同的安全组规则。

- True（默认）：同一网络中的主机节点允许所有类型的网络连接。在 Flat 网络中，允许所有租户内虚拟机实例的网络通信。在 VLAN 网络中，允许同一租户内虚拟机实例的网络通信。当然，也可以模拟该设置配置安全组，以允许虚拟机实例间的网络通信。
- False：对所有的网络通信，安全组规则强制执行。

安全组规则数量的限制由 security_group_rules 参数控制，安全组数量的限制由 security_groups 参数控制。

使用 nova 命令可以管理安全组的各项功能。

### 1. 查看安全组

（1）加载环境变量。

```
# source admin-openrc.sh
```

（2）列表显示所有的安全组。

```
# nova secgroup-list
+--------------------------------------+---------+------------------------+
| Id                                   | Name    | Description            |
+--------------------------------------+---------+------------------------+
| 005c8d5a-a436-4647-b785-9ec33a084ab3 | default | Default security group |
+--------------------------------------+---------+------------------------+
```

(3) 列表显示安全组内的所有规则。

```
# nova secgroup-list-rules default
+-------------+-----------+---------+-----------+--------------+
| IP Protocol | From Port | To Port | IP Range  | Source Group |
+-------------+-----------+---------+-----------+--------------+
| icmp        | -1        | -1      | 0.0.0.0/0 |              |
| tcp         | 22        | 22      | 0.0.0.0/0 |              |
|             |           |         |           | default      |
|             |           |         |           | default      |
+-------------+-----------+---------+-----------+--------------+
```

如上图所示，以上规则是允许的，其他规则默认是拒绝的。第一列显示的是网络协议，包括 ICMP、TCP 和 UDP；第二列和第三列显示的是端口范围；第四列显示的是 IP 地址范围。

### 2. 创建安全组

在创建安全组时，建议选择一个既简短又有明确意义的名称，该名称可以明确安全组的使用场景而无须过多解释。

(1) 加载环境变量。

```
# source admin-openrc.sh
```

(2) 创建安全组。

```
# nova secgroup-create GroupName Description
```

示例：

```
# nova secgroup-create global_http "Allows Web traffic anywhere on the Internet."
```

```
+--------------------------------------+-------------+----------------------------------------------+
| Id                                   | Name        | Description                                  |
+--------------------------------------+-------------+----------------------------------------------+
| 1578a08c-5139-4f3e-9012-86bd9dd9f23b | global_http | Allows Web traffic anywhere on the Internet. |
+--------------------------------------+-------------+----------------------------------------------+
```

(3) 添加安全组规则。

```
# nova secgroup-add-rule secGroupName ip-protocol from-port to-port CIDR
```

示例：

```
# nova secgroup-add-rule global_http tcp 80 80 0.0.0.0/0
```

```
+-------------+-----------+---------+-----------+--------------+
| IP Protocol | From Port | To Port | IP Range  | Source Group |
+-------------+-----------+---------+-----------+--------------+
| tcp         | 80        | 80      | 0.0.0.0/0 |              |
+-------------+-----------+---------+-----------+--------------+
```

(4) 查看规则是否设置成功。

```
# nova secgroup-list-rules global_http
+-------------+-----------+---------+-----------+--------------+
| IP Protocol | From Port | To Port | IP Range  | Source Group |
+-------------+-----------+---------+-----------+--------------+
| tcp         | 80        | 80      | 0.0.0.0/0 |              |
| tcp         | 443       | 443     | 0.0.0.0/0 |              |
+-------------+-----------+---------+-----------+--------------+
```

### 3. 删除安全组

(1) 加载环境变量。

```
# source admin-openrc.sh
```

(2) 删除安全组。

```
# nova secgroup-delete GroupName
```

示例：

```
# nova secgroup-delete global_http
```

## 21.4 管理主机集合

主机集合是管理员根据硬件资源的某一属性对硬件进行划分的功能，只对管理员可见。nova-scheduler 通过这一功能来进行虚拟机实例的调度。

### 1. 创建主机集合

同一主机节点可以被添加到多个主机集合中。添加主机节点到现存的主机集合中，需要使用编辑集合功能。

(1) 登录 dashboard。

(2) 在 Web 界面左侧，在 Admin（管理员）标签上，单击 System（系统）标签，选中

Host Aggregates(主机集合)分类,可以看到已定义的主机集合和可用域。

(3)单击 Create Host Aggregates(创建主机集合)按钮。

(4)在弹出的对话框中,填写或选择相应的值。

- Name(名称):主机集合的名称。

- Available Zone(可用域):OpenStack 项目中默认定义的可用域,包括 Nova 等。当然,也可以指定主机集合。

  ➢ 当主机集合作为可用域可见时,创建虚拟机实例时可以直接选择可用域。

  ➢ 当主机集合作为可用域不可见时,通过定义主机类型和其他参数来指定使用某主机集合。

(5)在 Manage Hosts within Aggregates(管理集合内的主机)标签中,单击"+"号将主机节点添加到集合中,如下图所示。

## 2. 编辑主机集合

(1)登录 dashboard。

(2)在 Web 界面左侧,在 Admin(管理员)标签上,单击 System(系统)标签,选中 Host Aggregates(主机集合)分类,可以看到已定义的主机集合和可用域。

（3）选择需要编辑的主机集合，单击 Edit Host Aggregate（编辑主机集合）按钮。

（4）在弹出的对话框中，可以修改主机集合的名字和可用域。

### 3．删除主机集合

（1）登录 dashboard。

（2）在 Web 界面左侧，在 Admin（管理员）标签上，单击 System（系统）标签，选中 Host Aggregates（主机集合）分类，可以看到已定义的主机集合和可用域。

（3）选择需要删除的主机集合，单击 Delete Host Aggregate（删除主机集合）按钮。

### 4．管理集合内主机

（1）登录 dashboard。

（2）在 Web 界面左侧，在 Admin（管理员）标签上，单击 System（系统）标签，选中 Host Aggregates（主机集合）分类，可以看到已定义的主机集合和可用域。

（3）选择需要管理的主机集合，单击"管理主机"按钮。

（4）在弹出的对话框中，单击"+"添加主机到集合中，单击"-"从集合中删除主机。

## 21.5 资源使用率统计

### 21.5.1 命令行方式

以下操作可以查询资源使用的基本统计信息。

> **注意：**
> 详细的资源使用信息可以通过 Ceilometer 计量服务组件提供的功能进行统计。

### 1．主机节点使用率统计

（1）列表展示主机节点及其上运行的 Nova 计算服务。

```
# nova host-list
```

```
+------------+-------------+----------+
| host_name  | service     | zone     |
+------------+-------------+----------+
| Controller | scheduler   | internal |
| Controller | conductor   | internal |
| Controller | consoleauth | internal |
| Controller | cert        | internal |
| Compute01  | compute     | nova     |
| Compute02  | compute     | nova     |
+------------+-------------+----------+
```

(2）主机节点资源使用率统计信息。

```
# nova host-describe Compute01
```

```
+-----------+----------------------------------+-----+-----------+---------+
| HOST      | PROJECT                          | cpu | memory_mb | disk_gb |
+-----------+----------------------------------+-----+-----------+---------+
| Compute01 | (total)                          | 4   | 3777      | 9       |
| Compute01 | (used_now)                       | 1   | 1024      | 1       |
| Compute01 | (used_max)                       | 1   | 512       | 1       |
| Compute01 | 9543b7db999641fda39ff53dc22e7bcb | 1   | 512       | 1       |
+-----------+----------------------------------+-----+-----------+---------+
```

cpu 列显示运行在该主机节点上的所有虚拟机实例消耗的 CPU 信息。

memory_mb 列显示运行在该主机节点上的所有虚拟机实例消耗的内存信息。

disk_gb 列显示运行在该主机节点上的所有虚拟机实例消耗的永久和临时存储信息。

PROJECT 列中的 used_now 行显示运行在该主机节点上的所有虚拟机实例消耗的资源与分配给主机本身虚拟机实例资源的总和。

PROJECT 列中的 used_max 行显示运行在该主机节点上的所有虚拟机实例消耗的资源信息。

 注意：

以上信息显示虚拟机实例的资源统计信息，并不能反映出主机节点的资源统计信息。

## 2. 实例使用率统计

（1）列表显示所有的实例信息。

```
# nova list
```

```
+--------------------------------------+------------+--------+------------+-------------+-------------------+
| ID                                   | Name       | Status | Task State | Power State | Networks          |
+--------------------------------------+------------+--------+------------+-------------+-------------------+
| 09bce5e1-0cad-4b19-a678-9cfe151237ff | instance01 | ACTIVE | -          | Running     | public=10.96.1.103|
| e4d633ef-9102-4974-b184-15a5d76539b1 | instance02 | ACTIVE | -          | Running     | public=10.96.1.105|
+--------------------------------------+------------+--------+------------+-------------+-------------------+
```

（2）实例诊断统计。

```
# nova diagnostics instance01
+------------------+----------------+
| Property         | Value          |
+------------------+----------------+
| vnet1_rx         | 1210744        |
| cpu0_time        | 19624610000000 |
| vda_read         | 0              |
| vda_write        | 0              |
| vda_write_req    | 0              |
| vnet1_tx         | 863734         |
| vnet1_tx_errors  | 0              |
| vnet1_rx_drop    | 0              |
| vnet1_tx_packets | 3855           |
| vnet1_tx_drop    | 0              |
| vnet1_rx_errors  | 0              |
| memory           | 2097152        |
| vnet1_rx_packets | 5485           |
| vda_read_req     | 0              |
| vda_errors       | -1             |
+------------------+----------------+
```

在默认情况下，仅管理员可以进行实例诊断统计。当然，也可以修改 policy.json 文件赋予其他用户权限。

（3）查询每个租户下的资源统计信息。

```
# nova usage-list
Usage from 2016-04-19 to 2016-05-18:
+----------------------------------+---------+--------------+-----------+--------------+
| Tenant ID                        | Servers | RAM MB-Hours | CPU Hours | Disk GB-Hours|
+----------------------------------+---------+--------------+-----------+--------------+
| 9543b7db999641fda39ff53dc22e7bcb | 2       | 286449.00    | 559.47    | 559.47       |
| efb33dccf25b40b399e92d6351fc7ea9 | 1       | 193949.31    | 378.81    | 378.81       |
+----------------------------------+---------+--------------+-----------+--------------+
```

## 21.5.2 图形界面方式

Ceilometer 计量服务组件提供了用户级别的资源使用率统计，为用户结算、系统监控和生产警告信息提供服务。数据通过 OpenStack 项目中的各服务组件发送的通知进行收集（如 Nova 计算服务组件等）或查询底层基础设施（如 Libvirt）。

**注意**：

仅管理员可以在 dashboard 上查看资源计量统计信息。

查看资源使用率统计信息的步骤如下。

（1）登录 dashboard。

（2）在 Web 界面左侧，在 Admin（管理员）标签上，单击 System（系统）标签，选中 Resource Usage（使用概况）分类。

（3）选择时间段，查看该租户下该时间段内的使用率报告，如下图所示。

## 21.6 查看系统服务信息

管理员可以在 dashboard 上查看关于 OpenStack 项目各服务组件的系统服务信息。

（1）登录 dashboard。

（2）在 Web 界面左侧，在 Admin（管理员）标签上，单击 System（系统）标签，选中 System Information（系统信息）分类。

（3）查看页面上 4 个标签的信息。

- Service（服务）标签显示已安装的 OpenStack 项目中的服务组件及其运行的主机节点名称和运行状态，如下图所示。

- Compute Service（计算服务）标签显示 Nova 计算服务组件的各项服务及其运行的主机节点、域和运行状态，如下图所示。

- Block Storage Service（块存储服务）标签显示 Cinder 块存储服务组件的各项服务及其运行的主机节点、域和运行状态，如下图所示。

- Network Agents（网络代理）标签显示各网络节点运行状态及其运行的主机节点，如下图所示。

另外，Orchestration Service（模板服务）标签显示 Heat 编排服务组件的各项服务及其运行状态。

# 第 22 章 仪表板使用

dashboard 的界面展示和功能排列可以修改。下面简单介绍 5 个功能的定制化：

- Logo（标志）。
- 图形界面。
- HTML 标题。
- Logo 链接。
- 帮助。

 注意：

本章对 dashboard 的定制化主要以/srv/www/openstack-dashboard/openstack_dashboard/local/local_settings.py 文件为例。

## 22.1 Logo 和图形界面定制化

（1）创建两张透明背景的 PNG 格式的图片，尺寸如下。

- 登录界面大小：365 像素 ×50 像素。

- 登录横幅大小：216 像素 ×35 像素。

（2）上传图片文件到/usr/share/openstack-dashboard/openstack_dashboard/static/dashboard/img/目录。

（3）在/usr/share/openstack-dashboard/openstack_dashboard/static/dashboard/scss/目录创建 CSS 文件。

（4）修改颜色和图片文件名。以下是 CSS 文件的例子。

```
/*
 * New theme colors for dashboard that override the defaults:
 *   dark blue: #355796 / rgb(53, 87, 150)
 *   light blue: #BAD3E1 / rgb(186, 211, 225)
 *
 * By Preston Lee <plee@tgen.org>
 */
h1.brand {
background: #355796 repeat-x top left;
border-bottom: 2px solid #BAD3E1;
}
h1.brand a {
background: url(../img/my_cloud_logo_small.png) top left no-repeat;
}
#splash .login {
background: #355796 url(../img/my_cloud_logo_medium.png) no-repeat center 35px;
}
#splash .login .modal-header {
border-top: 1px solid #BAD3E1;
}
.btn-primary {
background-image: none !important;
background-color: #355796 !important;
border: none !important;
box-shadow: none;
}
.btn-primary:hover,
.btn-primary:active {
```

```
    border: none;
    box-shadow: none;
    background-color: #BAD3E1 !important;
    text-decoration: none;
}
```

（5）编辑/usr/share/openstack-dashboard/openstack_dashboard/templates/_stylesheets.html 文件，添加以下内容。

```
<link href='{{ STATIC_URL }}bootstrap/css/bootstrap.min.css' media=
'screen' rel='stylesheet' />
  <link href='{{ STATIC_URL }}dashboard/css/{% choose_css %}' media='screen'
rel='stylesheet' />
  <link href='{{ STATIC_URL }}dashboard/css/custom.css' media='screen'
rel='stylesheet' />
```

（6）重启 Apache 服务。

（7）重新加载 dashboard 界面，检查设置是否生效。

## 22.2 HTML 标题、Logo 链接和帮助定制化

（1）设置 Web 界面窗口上方的 HTML 标题内容，编辑 local_settings.py 文件。

```
SITE_BRANDING = "Example, Inc. Cloud"
```

（2）重启 Apache 服务，使之生效。

（3）Logo 也可以是一个链接，编辑 local_settings.py 文件。

```
SITE_BRANDING_LINK = "http://example.com"
```

（4）重启 Apache 服务，使之生效。

（5）默认的帮助链接指向 http://docs.openstack.org，可以编辑 local_settings.py 文件修改帮助的指向 URL。

```
'help_url': "http://openstack.mycompany.org"
```

（6）重启 Apache 服务，使之生效。

# 第 23 章 管理镜像

管理员可以在具有权限的 project（租户）中创建镜像，供其他用户使用。

## 23.1 命令行方式

具有权限的用户可以上传和管理镜像，上传镜像使用 glance 命令，管理镜像使用 nova 命令。nova 命令可以查看和删除镜像，设置和删除 Image metadata（镜像元数据），以及使用镜像创建虚拟机实例等。

### 1. 查看镜像信息

```
# glance image-list
+--------------------------------------+--------+
| ID                                   | Name   |
+--------------------------------------+--------+
| 0ee987b9-4a8e-4908-8ab5-833b781f0502 | cirros |
+--------------------------------------+--------+

# glance image-show myCirrosImage
+------------------+--------------------------------------+
| Property         | Value                                |
+------------------+--------------------------------------+
| checksum         | ee1eca47dc88f4879d8a229cc70a07c6     |
| container_format | bare                                 |
| created_at       | 2016-02-23T15:04:07Z                 |
| disk_format      | qcow2                                |
| id               | 0ee987b9-4a8e-4908-8ab5-833b781f0502 |
| min_disk         | 0                                    |
```

```
| min_ram      | 0                                    |
| name         | cirros                               |
| owner        | efb33dccf25b40b399e92d6351fc7ea9     |
| protected    | False                                |
| size         | 13287936                             |
| status       | active                               |
| tags         | []                                   |
| updated_at   | 2016-02-23T15:04:08Z                 |
| virtual_size | None                                 |
| visibility   | public                               |
```

### 2. 创建和更新镜像

（1）创建镜像。

`# glance image-create imageName`

（2）更新镜像。

`# glance image-update imageName`

在创建和更新 Image 时可以使用以下属性。

- --architecture：选择操作系统位数。例如，i386 代表 32 位，x86_64 代表 64 位。

- --protected [True|False]：选择 True，仅具有权限的用户可以删除此镜像；反之亦然。

- --name <NAME>：镜像名字。

- --instance-uuid：用于创建镜像的实例 ID。

- --min-disk <MIN_DISK>：启动镜像所需的最小磁盘空间（GB）。

- --visibility <VISIBILITY>：定义 Image 可以使用的范围，VISIBILITY 值为 public 或 private。

- --kernel-id <KERNEL_ID>：在 Glance 中存储的 Image ID，当启动 AMI-style Image 时作为 Kernel 被使用。

- --os-version <OS_VERSION>：操作系统版本。

- --disk-format <DISK_FORMAT>：镜像格式，有效的格式包括 ami、ari、aki、vhd、vmdk、raw、qcow2、vdi 和 iso。

- --id <ID>：Image 的唯一 ID。

- --owner <OWNER>：Image 的属主。

- --ramdisk-id <RAMDISK_ID>：在 Glance 中存储的 Image ID，当启动 AMI-style Image 时作为 Ramdisk 被使用。

- --min-ram <MIN_RAM>：启动镜像所需的最小内存（MB）。

- --container-format <CONTAINER_FORMAT>：容器的格式，有效的格式包括 ami、ari、aki、bare、ovf 和 ova。

- --property <key=value>：定义 Image 的任意属性值，允许使用多次。

- --file <FILE>：在创建过程中上传的本地 Image 文件。

- --progress：显示上传 Image 文件的进度。

以下举例说明如何上传镜像文件。上传文件为 CirrOS Image，格式为 qcow2，并设置为允许所有人访问。

```
# glance image-create --name cirros --disk-format qcow2 \
  --container-format bare --is-public True --file \
 /var/lib/glance/images/cirros-0.3.4-x86_64-disk.img
```

以下举例说明如何更新 Image 文件。定义 Image 的 Disk Bus（磁盘总线）驱动、CD-ROM（光驱）驱动和 VIF（网卡）驱动。

```
# glance image-update \
   --property hw_disk_bus=scsi \
   --property hw_cdrom_bus=ide \
   --property hw_vif_model=e1000 \
   f16-x86_64-openstack-sda
```

基于使用的 Hypervisor 类型（在配置文件/etc/nova/nova.conf 中定义）决定了 Libvirt 虚拟化工具所使用的 Disk、CD-ROM 和 VIF 的驱动类型。为了获得最佳性能，Libvirt 默认使用 Virtio 作为磁盘总线和网卡驱动类型。这种方式的缺点是缺少 Virtio 驱动的操作系统不能运行，如 BSD、Solaris、版本比较老的 Linux 和 Windows。

磁盘和光驱的兼容性列表如下表所示。

| Hypervisor 类型 | 支持驱动类型 |
| --- | --- |
| KVM 或 QEMU | IDE,SCSI,Virtio |
| Xen | IDE,Xen |

网卡的兼容性列表如下表所示。

| Hypervisor 类型 | 支持驱动类型 |
| --- | --- |
| KVM 或 QEMU | E1000,ne2k_pci,PCNet,rtl8139,Virtio |
| Xen | E1000,netfront,pcnet,rtl8139,ne2k_pci |
| VMware | VirtualE1000, VirtualPCNet32, VirtualVmxnet |

### 3. 错误排查

下列信息可以帮助解决创建和管理 Image 过程中遇到的问题。

- 检查正在使用的 QEMU 的版本是否大于或等于 0.14。如果不是，则在 nova-compute.log 文件中会出现错误信息。
- 在/var/log/nova/nova-api.log 和/var/log/nova/nova-compute.log 文件中查找错误信息。

## 23.2 图形界面方式

### 1. 创建镜像

（1）使用 admin 用户登录 dashboard。

（2）在 Web 界面左侧，打开 System（系统）标签，选中 Images（镜像）分类，可以看到管理的所有镜像。

（3）单击 Create Image（创建镜像）按钮，进入如下图所示的界面。

(4) 在上述界面中选择和填写如下表所示的信息。

| Name | 镜像名字 |
| --- | --- |
| Description | 对镜像的描述信息 |
| Image Source | 在下拉列表中选择镜像来源,包括镜像文件和镜像位置两个选项 |
| Image File or Image Location | 基于用户的选择。镜像文件存在当前的 PC 磁盘上,镜像位置包含在 URL 中 |
| Format | 镜像的格式 |
| Architecture | 选择操作系统位数。例如,i386 代表 32 位,x86_64 代表 64 位 |
| Minimum Disk (GB) | 启动镜像所需的最小磁盘空间。可为空 |
| Minimum RAM (MB) | 启动镜像所需的最小内存。可为空 |
| Copy Data | 选择此项,复制镜像数据到 Image 服务;不选,使用当前位置的数据 |
| Public | 选择此项,所有用户都可以使用此镜像 |
| Protected | 选择此项,仅有权限用户可以删除此镜像 |

(5)单击 Create Image(创建镜像)按钮。镜像开始上传,一段时间后,状态由 QUEUED(排队)变为 ACTIVE(激活)。

## 2. 更新镜像

(1)使用 admin 用户登录 dashboard。

(2)在 Web 界面左侧,打开 System(系统)标签,选中 Images(镜像)分类。

(3)选择需要编辑的镜像,单击 Edit Image(编辑镜像)按钮。

(4)在弹出的"更新镜像"窗口中修改需要更新的项,如镜像的名字、镜像的格式等。

(5)单击 Update Image(更新镜像)按钮保存镜像。

## 3. 删除镜像

(1)使用 admin 用户登录 dashboard。

(2)在 Web 界面左侧,打开 System(系统)标签,选中 Images(镜像)分类。

① 选择需要删除的 Image。

② 单击 Delete Images(删除镜像)按钮。

③ 在弹出的确认对话框中,确认删除。

警告:

此操作不能回退。

# 第 24 章 管理网络

OpenStack 项目中的 Neutron 网络服务组件是一个可横向扩展的组件,管理着 OpenStack 环境中的所有网络连接,包括创建网络、子网、端口和分配 IP 地址等。

## 24.1 命令行方式

### 1. 创建指定类型的网络

```
# neutron net-create public --shared --provider:physical_network public
--provider:network_type flat
+---------------------------+--------------------------------------+
| Field                     | Value                                |
+---------------------------+--------------------------------------+
| admin_state_up            | True                                 |
| id                        | 37193c68-b70c-4153-b5cf-b6f19e9d8951 |
| mtu                       | 0                                    |
| name                      | public                               |
| port_security_enabled     | True                                 |
| provider:network_type     | flat                                 |
| provider:physical_network | public                               |
| provider:segmentation_id  |                                      |
| router:external           | False                                |
| shared                    | True                                 |
| status                    | ACTIVE                               |
| subnets                   | b51b9bdb-05d8-40c5-aa73-c70557bc46a9 |
|                           | a60f0452-ba3d-41bd-83d7-86a38c12f5b5 |
| tenant_id                 | efb33dccf25b40b399e92d6351fc7ea9     |
+---------------------------+--------------------------------------+
```

以上命令中的--provider:network-type 选项指定 Neutron 网络服务组件使用 flat 类型的网络拓扑，也可以选择 vlan、local 等类型的拓扑。

### 2. 创建子网

```
# neutron subnet-create public 1.1.11.0/24 --name public_01 \
  --allocation-pool   start=1.1.11.100,end=1.1.11.200   --dns-nameserver
8.8.4.4 --gateway 1.1.11.1
```

```
+-------------------+------------------------------------------------+
| Field             | Value                                          |
+-------------------+------------------------------------------------+
| allocation_pools  | {"start": "1.1.11.100", "end": "1.1.11.200"}   |
| cidr              | 1.1.11.0/24                                    |
| dns_nameservers   | 8.8.4.4                                        |
| enable_dhcp       | True                                           |
| gateway_ip        | 1.1.11.1                                       |
| host_routes       |                                                |
| id                | a60f0452-ba3d-41bd-83d7-86a38c12f5b5            |
| ip_version        | 4                                              |
| ipv6_address_mode |                                                |
| ipv6_ra_mode      |                                                |
| name              | public_01                                      |
| network_id        | 37193c68-b70c-4153-b5cf-b6f19e9d8951            |
| subnetpool_id     |                                                |
| tenant_id         | efb33dccf25b40b399e92d6351fc7ea9               |
+-------------------+------------------------------------------------+
```

neutron subnet-create 命令包括以下必选和可选参数。

- 指定该子网所属的网络名称或 ID：在本例中，public 是创建子网所属的网络名称。

- 指定子网的范围区间：在本例中，--allocation-pool start=1.1.11.100,end=1.1.11.200 是创建子网的范围区间。

- 指定子网名称：在本例中，public_01 是创建子网的名称。

- 指定子网所使用的 DNS 服务器：在本例中，8.8.4.4 是创建子网所使用的 DNS 服务器。

- 指定子网所使用的网关：在本例中，1.1.11.1 是创建子网所使用的网关。

### 3. 创建端口

（1）使用指定 IP 地址创建端口。

```
# neutron port-create public --fixed-ip ip_address=10.96.1.103
```

```
+------------------------+---------------------------------------------------------------------------------------------+
| Field                  | Value                                                                                       |
+------------------------+---------------------------------------------------------------------------------------------+
| admin_state_up         | True                                                                                        |
| allowed_address_pairs  |                                                                                             |
| binding:host_id        | Compute01                                                                                   |
| binding:profile        | {}                                                                                          |
| binding:vif_details    | {"port_filter": true}                                                                       |
| binding:vif_type       | bridge                                                                                      |
| binding:vnic_type      | normal                                                                                      |
| device_id              | 09bce5e1-0cad-4b19-a678-9cfe151237ff                                                        |
| device_owner           | compute:None                                                                                |
| dns_assignment         | {"hostname": "host-10-96-1-103", "ip_address": "10.96.1.103", "fqdn": "host-10-96-1-103.openstacklocal."} |
| dns_name               |                                                                                             |
| extra_dhcp_opts        |                                                                                             |
| fixed_ips              | {"subnet_id": "b51b9bdb-05d8-40c5-aa73-c70557bc46a9", "ip_address": "10.96.1.103"}          |
| id                     | 81aaa69b-abc0-493b-a737-3d82f7bea729                                                        |
| mac_address            | fa:16:3e:40:5e:0f                                                                           |
| name                   |                                                                                             |
| network_id             | 37193c68-b70c-4153-b5cf-b6f19e9d8951                                                        |
| port_security_enabled  | True                                                                                        |
| security_groups        | e4ad703c-f8de-44cd-bcc6-d111da6756fe                                                        |
| status                 | ACTIVE                                                                                      |
| tenant_id              | 9543b7db999641fda39ff53dc22e7bcb                                                            |
+------------------------+---------------------------------------------------------------------------------------------+
```

neutron port-create 命令中的 public 是网络名称，是必选参数；--fixed-ip ip_address=10.96.1.103 是可选参数，用于指定所需的 IP 地址。

（2）创建端口。

```
# neutron port-create public
```

```
+------------------------+---------------------------------------------------------------------------------------------+
| Field                  | Value                                                                                       |
+------------------------+---------------------------------------------------------------------------------------------+
| admin_state_up         | True                                                                                        |
| allowed_address_pairs  |                                                                                             |
| binding:host_id        | Compute02                                                                                   |
| binding:profile        | {}                                                                                          |
| binding:vif_details    | {"port_filter": true}                                                                       |
| binding:vif_type       | bridge                                                                                      |
| binding:vnic_type      | normal                                                                                      |
| device_id              | e4d633ef-9102-4974-b184-15a5d76539b1                                                        |
| device_owner           | compute:None                                                                                |
| dns_assignment         | {"hostname": "host-10-96-1-105", "ip_address": "10.96.1.105", "fqdn": "host-10-96-1-105.openstacklocal."} |
| dns_name               |                                                                                             |
| extra_dhcp_opts        |                                                                                             |
| fixed_ips              | {"subnet_id": "b51b9bdb-05d8-40c5-aa73-c70557bc46a9", "ip_address": "10.96.1.105"}          |
| id                     | 9beae6d1-4b1f-437d-9559-af9db951abf1                                                        |
| mac_address            | fa:16:3e:4b:7c:b6                                                                           |
| name                   |                                                                                             |
| network_id             | 37193c68-b70c-4153-b5cf-b6f19e9d8951                                                        |
| port_security_enabled  | True                                                                                        |
| security_groups        | e4ad703c-f8de-44cd-bcc6-d111da6756fe                                                        |
| status                 | ACTIVE                                                                                      |
| tenant_id              | 9543b7db999641fda39ff53dc22e7bcb                                                            |
+------------------------+---------------------------------------------------------------------------------------------+
```

在没有指定 IP 地址的情况下，neutron port-create 命令会随机为该端口生成一个 IP 地址。

（3）查看端口列表。

```
# neutron port-list
```

```
+--------------------------------------+------+-------------------+----------------------+
| id                                   | name | mac_address       | fixed_ips            |
+--------------------------------------+------+-------------------+----------------------+
| 6a946805-3e5b-4fdc-8a41-f9e7d0865c62 |      | fa:16:3e:fa:07:34 | {"subnet_id": "      |
|                                      |      |                   | {"subnet_id": "      |
| 81aaa69b-abc0-493b-a737-3d82f7bea729 |      | fa:16:3e:40:5e:0f | {"subnet_id": "      |
| 9beae6d1-4b1f-437d-9559-af9db951abf1 |      | fa:16:3e:4b:7c:b6 | {"subnet_id": "      |
+--------------------------------------+------+-------------------+----------------------+
```

## 24.2 图形界面方式

在 dashboard 图形界面方式中,可以进行创建网络、路由和端口等操作。本节简单介绍如何创建网络和端口,具体如下。

### 1. 创建网络

(1) 登录 dashboard。

(2) 在 Web 界面左侧,在 Project(项目)标签上,单击 Network(网络)标签,选中 Networks(网络)选项。

(3) 单击 Create Network(创建网络)按钮。

(4) 在创建网络的弹出框中,需要指定以下值。

① Network(网络)标签。

- Network Name(网络):需创建的网络名称。

- Admin State(管理员状态):状态是自动网络。

- Create Subnet(创建子网):选择该项,创建子网。如果不选择创建子网,则虚拟机实例无法使用该网络。

② Subnet(子网)标签。

- Subnet Name(子网名称):需创建的子网名称。

- Network Address(网络地址):需创建子网的 IP 地址范围。

- IP Version(IP 版本):选择 IPv4 或 IPv6。

- Gateway IP(网关 IP):指定网关 IP 地址,可选。

- Disable Gateway(禁用网关):选择该项,禁用网关。

③ Subnet Details(子网详情)标签。

- Enable DHCP(激活 DHCP):选择该项,启用 DHCP 功能。

- Allocation Pools（分配地址池）：可以分配的 IP 地址范围。
- DNS Name Servers（DNS 域名解析服务）：指定 DNS 服务器地址。
- Host Routes（主机路由）：为主机路由指定 IP 地址。

（5）单击 Create（创建）按钮，创建网络，如下图所示。

## 2. 创建端口

（1）登录 dashboard。

（2）在 Web 界面左侧，在 Project（项目）标签上，单击 Network（网络）标签，选中 Networks（网络）选项。

（3）选择需要在网络上创建端口的网络名称。

（4）在 Create Port（创建端口）对话框中，指定以下值。

- Name（名称）：需创建的端口名称。
- Device ID（设备 ID）：分配给端口使用的设备 ID。
- Device Owner（设备属主）：分配给端口使用的设备属主。
- Binding Host（绑定主机）：需创建端口所在的主机节点 ID。
- Binding VNIC Type（绑定类型）：指定虚拟端口类型。

（5）单击 Create Port（创建端口）按钮，创建端口。

# 第 25 章 管理卷设备

卷设备是一种可拆卸的块存储设备（Block Storage），类似于可插拔的 USB 设备。同一个卷设备只能分配给一个虚拟机实例作为持久性存储使用。虚拟机实例上的卷设备可以随时分配或解除使用。

## 25.1 命令行方式

在命令行操作中，nova 和 cinder 命令组合使用，共同管理卷设备。

### 1. 迁移卷设备

管理员可以在存储节点间迁移卷设备，而不破坏卷设备上的数据。卷设备迁移只适用于没有使用且没有 snapshot 的卷设备。

使用案例：

- 减少存储节点的维护时间。
- 修改卷设备的属性。

- 释放空间。

```
# cinder migrate volumeID destinationHost --force-host-copy True|False
```

其中，--force-host-copy True 属性强制使用通用的卷设备迁移策略，忽略任何驱动优化。

 注意：

　　如果卷设备正在被使用或者含有 snapshot，则指定的迁移目标主机不会接收该卷设备。非管理员用户不能执行卷迁移操作。

### 2. 创建卷设备

本例展示如何创建一个基于 Image 的卷设备。

（1）列表展示镜像。

```
# nova image-list
```

（2）列表展示 Zone。

```
# cinder availability-zone-list
```

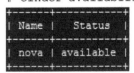

（3）在 Nova Zone 中创建基于 CirrOS Image 的卷设备。

```
# cinder create 1 --display-name boot_volume01_cirros --image-id 0ee987b9-4a8e-4908-8ab5-833b781f0502 --availability-zone nova
```

```
attachments              | [{u'server_id': u'e4d633ef-9102-4974-b184-15a5d76539b1', u'attachment_id': u'c3ec024c-aab1-4136-8495-a15e7fb6c00a', u'host_name': None, u'volume_id': u'a8a1c8a1-d046-4c08-a9fa-6ad6860e0ea8', u'device': u'/dev/vda', u'id': u'a8a1c8a1-d046-4c08-a9fa-6ad6860e0ea8'}]
availability_zone        | nova
bootable                 | true
consistencygroup_id      | None
created_at               | 2016-05-06T06:17:34.000000
description              |
encrypted                | False
id                       | a8a1c8a1-d046-4c08-a9fa-6ad6860e0ea8
metadata                 | {u'readonly': u'False', u'attached_mode': u'rw'}
multiattach              | False
name                     | boot_volume01_cirros
os-vol-tenant-attr:tenant_id | 9543b7db999641fda39ff53dc22e7bcb
os-volume-replication:driver_data | None
os-volume-replication:extended_status | None
replication_status       | disabled
size                     | 1
snapshot_id              | None
source_volid             | None
status                   | in-use
user_id                  | 33cd0cb0e80a4ba6bb5b8c36947f6267
volume_image_metadata    | {u'container_format': u'bare', u'min_ram': u'0', u'disk_format': u'qcow2', u'image_name': u'cirros', u'image_id': u'0ee987b9-4a8e-4908-8ab5-833b781f0502', u'checksum': u'eeleca47dc08f4879d2a229cc70a07c6', u'min_disk': u'0', u'size': u'13287936'}
volume_type              | None
```

（4）检查卷设备是否成功创建。

```
# cinder list
+--------------------------------------+--------+----------------------+------+-------------+----------+-------------+--------------------------------------+
|                  ID                  | Status |         Name         | Size | Volume Type | Bootable | Multiattach |             Attached to              |
+--------------------------------------+--------+----------------------+------+-------------+----------+-------------+--------------------------------------+
| 81e14efa-cc7c-4837-8763-d450a63a93ea | in-use |       volume01       |  1   |      -      |  false   |    False    | 09bce5e1-0cad-4b19-a678-9cfe151237ff |
| a8a1c8a1-d046-4c08-a9fa-6ad6860e0ea8 | in-use | boot_volume01_cirros |  1   |      -      |   true   |    False    | e4d633ef-9102-4974-b184-15a5d76539b1 |
+--------------------------------------+--------+----------------------+------+-------------+----------+-------------+--------------------------------------+
```

创建成功的卷设备状态（Status）是 available。

### 3. 创建特定类型的卷设备

Cinder 支持以下三种类型的卷设备。

- 普通的卷设备类型。

```
# cinder create --name <volume name> --volume-type <volume type> <size>
```

- 镜像指定的卷设备类型（cinder_img_volume_type 属性）。

当镜像含有 cinder_img_volume_type 属性时，Cinder 创建卷设备时会默认使用该卷设备类型。

```
# glance image-list
+--------------------------------------+--------+
| ID                                   | Name   |
+--------------------------------------+--------+
| 0ee987b9-4a8e-4908-8ab5-833b781f0502 | cirros |
+--------------------------------------+--------+
```

```
# glance image-show 0ee987b9-4a8e-4908-8ab5-833b781f0502
+----------------------------+--------------------------------------+
| Property                   | Value                                |
+----------------------------+--------------------------------------+
| checksum                   | ee1eca47dc88f4879d8a229cc70a07c6     |
| cinder_img_volume_type     | lvmdriver-1                          |
| container_format           | ami                                  |
| created_at                 | 2016-02-23T15:04:07Z                 |
| disk_format                | ami                                  |
| id                         | 0ee987b9-4a8e-4908-8ab5-833b781f0502 |
| kernel_id                  | 6cf01154-0408-416a-b69c-b28b48c5dt5f |
| min_disk                   | 0                                    |
| min_ram                    | 0                                    |
| name                       | cirros                               |
| owner                      | efb33dccf25b40b399e92d6351fc7ea9     |
| protected                  | False                                |
| ramdisk_id                 | de457c7c-2038-435d-abed-5dfa6430e5fe |
| size                       | 13287936                             |
| status                     | active                               |
| tags                       | []                                   |
| updated_at                 | 2016-02-23T15:04:08Z                 |
| virtual_size               | None                                 |
| visibility                 | public                               |
+----------------------------+--------------------------------------+
```

# cinder create --name test --image-id 0ee987b9-4a8e-4908-8ab5-833b781f0502 1

```
+---------------------------------------+--------------------------------------+
| Property                              | Value                                |
+---------------------------------------+--------------------------------------+
| attachments                           | []                                   |
| availability_zone                     | nova                                 |
| bootable                              | false                                |
| consistencygroup_id                   | None                                 |
| created_at                            | 2016-02-23T23:17:51.000000           |
| description                           | None                                 |
| encrypted                             | False                                |
| id                                    | a8a1c8a1-d046-4c08-a9fa-6ad6880e0ea8 |
| metadata                              | {}                                   |
| migration_status                      | None                                 |
| multiattach                           | False                                |
| name                                  | boot_volume01_cirros                 |
| os-vol-host-attr:host                 | None                                 |
| os-vol-mig-status-attr:migstat        | None                                 |
| os-vol-mig-status-attr:name_id        | None                                 |
| os-vol-tenant-attr:tenant_id          | 4c0dbc92040c41b1bdb3827653682952     |
| os-volume-replication:driver_data     | None                                 |
| os-volume-replication:extended_status | None                                 |
| replication_status                    | disabled                             |
| size                                  | 1                                    |
| snapshot_id                           | None                                 |
| source_volid                          | None                                 |
| status                                | creating                             |
| updated_at                            | None                                 |
| user_id                               | 9a125f3d111e47e6a25f573853b32fd9     |
| volume_type                           | lvmdriver-1                          |
+---------------------------------------+--------------------------------------+
```

- 默认的卷设备类型（在 cinder.conf 中定义）。

如果没有使用上述两种方式设定卷设备类型，则默认使用在配置文件中定义的类型。

```
[default]
default_volume_type = lvmdriver-1
```

## 4. 分配卷设备

（1）指定卷设备 ID 和虚拟机实例 ID，将卷设备分配给虚拟机实例。

```
# nova volume-attach 09bce5e1-0cad-4b19-a678-9cfe151237ff 81e14efa-cc7c-4837-8763-d450a63a93ea /dev/vdb
```

```
+----------+--------------------------------------+
| Property | Value                                |
+----------+--------------------------------------+
| device   | /dev/vdb                             |
| serverId | 09bce5e1-0cad-4b19-a678-9cfe151237ff |
| id       | 81e14efa-cc7c-4837-8763-d450a63a93ea |
| volumeId | 81e14efa-cc7c-4837-8763-d450a63a93ea |
+----------+--------------------------------------+
```

（2）显示卷设备信息。

```
# cinder show 81e14efa-cc7c-4837-8763-d450a63a93ea
```

## 5. 调整卷设备容量

（1）调整卷设备容量，首先需要分离虚拟机实例与卷设备的映射。

```
# nova volume-detach 09bce5e1-0cad-4b19-a678-9cfe151237ff 81e14efa-cc7c-4837-8763-d450a63a93ea
```

volume-detach 命令没有输出。

（2）列表展示卷设备。

```
# cinder list
```

| ID | Status | Name | Size | Volume Type | Bootable | Multiattach | Attached to |
| --- | --- | --- | --- | --- | --- | --- | --- |
| 81e14efa-cc7c-4837-8763-d450a63a93ea | available | volume01 | 1 | - | false | False | |
| a8a1c8a1-d046-4c05-a9fa-6ad6880e0ea8 | in-use | boot_volume01_cirros | 1 | - | true | False | e4d633ef-9102-4974-b184-15a5d76539b1 |

卷设备状态由 in-use 转变为 available。

（3）调整卷设备容量大小。

```
# cinder extend 81e14efa-cc7c-4837-8763-d450a63a93ea 8
```

cinder extend 命令没有输出。

## 6. 删除卷设备

（1）分离虚拟机实例与卷设备的映射。

```
# nova volume-detach 09bce5e1-0cad-4b19-a678-9cfe151237ff 81e14efa-
cc7c-4837-8763-d450a63a93ea
```

volume-detach 命令没有输出。

(2) 列表展示卷设备。

```
# cinder list
```

| ID | Status | Name | Size | Volume Type | Bootable | Multiattach | Attached to |
| --- | --- | --- | --- | --- | --- | --- | --- |
| 81e14efa-cc7c-4837-8763-d450a63a93ea | available | volume01 | 1 | - | false | False | |
| a8a1c8a1-d046-4c08-a9fa-6ad6880e0ea8 | in-use | boot_volume01_cirros | 1 | - | true | False | e4d633ef-9102-4974-b184-15a5d76539b1 |

卷设备状态由 in-use 转变为 available。

(3) 删除卷设备。

```
# cinder delete volume01
```

(4) 检查被删除卷设备的状态。

```
# cinder list
```

| ID | Status | Name | Size | Volume Type | Bootable | Multiattach | Attached to |
| --- | --- | --- | --- | --- | --- | --- | --- |
| 81e14efa-cc7c-4837-8763-d450a63a93ea | deleting | volume01 | 1 | - | false | False | |
| a8a1c8a1-d046-4c08-a9fa-6ad6880e0ea8 | in-use | boot_volume01_cirros | 1 | - | true | False | e4d633ef-9102-4974-b184-15a5d76539b1 |

## 7. 传送卷设备

卷设备属主创建传送请求，同时将传送卷设备 ID 和授权 key 发送给接收卷设备的用户，接收卷设备的用户根据传送卷 ID 和授权 key 接收卷设备。传送卷操作只能在同一个 OpenStack 项目下的租户内进行，传送卷用户和接收卷用户必须在同一租户内。

使用案例：

- 创建自定义的启动卷设备或含有应用数据的卷设备并传送给用户。
- 数据的批量导入。

下面举例说明如何进行卷设备传送。

(1) 列表展示所有卷设备。

```
# cinder list
```

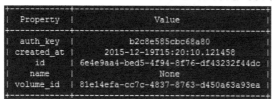

（2）创建卷设备传送请求。

```
# cinder transfer-create 81e14efa-cc7c-4837-8763-d450a63a93ea
```

```
+-------------+--------------------------------------+
|  Property   |                Value                 |
+-------------+--------------------------------------+
|  auth_key   |           b2c8e585cbc68a80           |
| created_at  |      2015-12-19T15:20:10.121458      |
|     id      | 6e4e9aa4-bed5-4f94-8f76-df43232f44dc |
|    name     |                 None                 |
|  volume_id  | 81e14efa-cc7c-4837-8763-d450a63a93ea |
+-------------+--------------------------------------+
```

传送的卷设备状态必须是 available，否则请求会被拒绝。该请求成功操作时，卷设备的状态为 awaiting-transfer。

 注意：

通过指定 --display-name displayName 参数可以为传送卷定义名称。

 注意：

创建传送卷 cinder transfer-create 时 auth_key 是可见的，之后执行 cinder transfer-show TRANSFER_ID 时 auth_key 已经不可见。

（3）查看等待传送的卷设备。

```
# cinder transfer-list
```

```
+--------------------------------------+--------------------------------------+------+
|                  ID                  |              VolumeID                | Name |
+--------------------------------------+--------------------------------------+------+
| 6e4e9aa4-bed5-4f94-8f76-df43232f44dc | 81e14efa-cc7c-4837-8763-d450a63a93ea | None |
+--------------------------------------+--------------------------------------+------+
```

（4）卷设备被成功传送后，等待传送的卷设备在队列中消失。

```
# cinder transfer-list
```

（5）接收被传送的卷设备。

```
# cinder transfer-accept 6e4e9aa4-bed5-4f94-8f76-df43232f44dc b2c8e585cbc68a80
```

```
+--------------+-----------------------------------------+
|   Property   |                  Value                  |
+--------------+-----------------------------------------+
|      id      |  6e4e9aa4-bed5-4f94-8f76-df43232f44dc   |
|     name     |                  None                   |
|  volume_id   |  81e14efa-cc7c-4837-8763-d450a63a93ea   |
+--------------+-----------------------------------------+
```

接收被传送的卷设备时，需要指定 transferID 和 authKey。

## 8. 删除传送卷设备

（1）列表展示卷设备的状态。

```
# cinder list
+-----------+------------------+--------------+------+-------------+----------+-------------+
|    ID     |      Status      | Display Name | Size | Volume Type | Bootable | Attached to |
+-----------+------------------+--------------+------+-------------+----------+-------------+
| 72bfce9f..|      error       |     None     |   1  |     None    |  false   |             |
| 81e14efa..| awaiting-transfer|     None     |   1  |     None    |  false   |             |
+-----------+------------------+--------------+------+-------------+----------+-------------+
```

（2）列表展示传送卷设备。

```
# cinder transfer-list
+--------------------------------------+--------------------------------------+------+
|                  ID                  |               VolumeID               | Name |
+--------------------------------------+--------------------------------------+------+
| 6e4e9aa4-bed5-4f94-8f76-df43232f44dc | 81e14efa-cc7c-4837-8763-d450a63a93ea | None |
+--------------------------------------+--------------------------------------+------+
```

（3）删除正在传送卷设备。

```
# cinder transfer-delete transferID
```

（4）检查传送卷是否被成功删除。

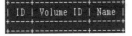

```
# cinder list
+--------------------------------------+----------+--------------------+------+-------------+----------+-------------+--------------------------------------+
|                  ID                  |  Status  |        Name        | Size | Volume Type | Bootable | Multiattach |             Attached to              |
+--------------------------------------+----------+--------------------+------+-------------+----------+-------------+--------------------------------------+
| 81e14efa-cc7c-4837-8763-d450a63a93ea | available|      volume01      |   1  |      -      |  false   |    False    |                                      |
| a5a1c8a1-d046-4c08-a9fa-6ad6880e0ea8 |  in-use  | boot_volume01_cirros|  1  |      -      |  true    |    False    | e4d633ef-9102-4974-b184-15a5d76539b1 |
+--------------------------------------+----------+--------------------+------+-------------+----------+-------------+--------------------------------------+
```

## 25.2 图形界面方式

用户可以在租户内管理卷设备和卷设备类型，执行创建和删除操作，随时将卷设备从虚拟机实例上进行挂载或分离。同时，卷设备支持进行加密操作。

### 1. 创建卷设备类型

（1）登录 dashboard。

（2）在 Web 界面左侧，在 Admin（管理员）标签上，单击 System（系统）标签，选中 Volume（云硬盘）分类，可以看到管理的所有卷设备信息。

（3）选择 Volume Types（云硬盘类型）选项，单击 Create Volume Type（创建云硬盘类型）按钮，输入卷设备类型名称。

（4）单击 Create Volume Type（创建云硬盘类型）按钮，进行创建。

注意：

界面出现信息提示操作是否成功。

### 2. 创建加密卷设备类型

（1）首先按照步骤创建卷设备类型（参考上述步骤）。

（2）在卷设备类型右方选择 Create Encryption（创建加密）。

（3）在弹出的对话框中填写如下内容，如下图所示。

- Provider（提供者）：指定加密类型。
- Control Location（控制地点）：指定加密是由前端控制（Nova）还是由后端控制（Cinder）。
- Cipher：指定加密算法。
- Key Size（密钥大小）：指定密钥大小。

加密卷设备类型功能各选项的有效值如下表所示。

| | |
|---|---|
| Provider（提供者） | nova.volume.encryptors. luks.LuksEncryptor（建议） |
| | nova.volume.encryptors. cryptsetup.CryptsetupEncryptor |
| Control Location（控制地点） | front-end（建议） |
| | back-end |
| Cipher | aes-xts-plain64（建议） |
| | aes-cbc-essiv |
| Key Size（密钥大小） | aes-xts-plain64:512 |
| | aes-cbc-essiv:256 |
| | 256 |

## 3. 删除卷设备类型

卷设备类型被删除时，属于该卷设备类型的卷设备不会被删除。

（1）登录 dashboard。

（2）在 Web 界面左侧，在 Admin（管理员）标签上，单击 System（系统）标签，选中 Volume（云硬盘）分类，可以看到管理的所有卷设备信息。

（3）选择 Volume Types（云硬盘类型）选项，选择需要删除的卷设备类型，单击 Delete Volume Type（删除云硬盘类型）按钮删除。

（4）在弹出的对话框中，确认删除。

注意：
界面出现信息提示操作是否成功。

### 4．删除卷设备

虚拟机实例被删除后，挂载在该虚拟机实例上的卷设备数据不会受到影响。

（1）登录 dashboard。

（2）在 Web 界面左侧，在 Admin（管理员）标签上，单击 System（系统）标签，选中 Volume（云硬盘）分类，可以看到管理的所有卷设备信息。

（3）选择需要删除的卷设备，单击 Delete Volume（删除云硬盘）按钮。

（4）在弹出的对话框中，确认删除。

注意：
界面出现信息提示操作是否成功。

# 第 26 章 管理虚拟机实例

本章具体介绍对虚拟机实例进行管理的几个重要案例。之前章节的介绍是针对虚拟机实例的某个操作,没有从多方面综合介绍,而本章从多个方面对虚拟机实例的综合管理进行介绍,具有实践意义。

## 26.1 创建虚拟机实例

虚拟机实例运行在 OpenStack 项目中,为用户提供各种计算资源,包括 CPU、内存和硬盘等。

### 26.1.1 命令行方式

在创建虚拟机实例之前,首先需要确定以下信息。

- 启动源:启动源可以是镜像、快照或包含镜像或快照的块存储设备。
- 主机类型(Flavor):主机实例就是服务器的硬件配置清单,定义了虚拟机实例可以使用的硬件资源,包括 CPU、内存数量和磁盘容量。

- 密钥对（Key Pair）：密钥对是一个 SSH 凭证，当虚拟机实例启动时写入镜像中并加载该密钥对。

- 安全组（Security Group）：安全组定义防火墙策略，控制流向虚拟机实例的网络流量。

下面通过三种不同的启动源介绍如何创建虚拟机实例。

### 1. 通过镜像创建虚拟机实例

当创建虚拟机实例的必需信息（主机类型、密钥对、安全组等）确定后，可以使用以下 nova boot 命令创建虚拟机实例。

```
# nova boot --flavor FLAVOR_ID --image IMAGE_ID --key-name KEY_NAME \
  --user-data USER_DATA_FILE --security-groups SEC_GROUP_NAME --meta KEY=VALUE INSTANCE_NAME
```

虚拟机实例名称 INSTANCE_NAME 不允许大于 63 个字符，否则 dnsmasq 不能正常工作。为了解决此问题，Nova 计算服务组件会自动截取字符串作为主机名，相关的警告信息会打印在 nova-network.log 日志文件中。

 注意：

在没有提供或提供不合适的密钥对、安全组等信息时，访问虚拟机实例只能通过 VNC，远程访问是被拒绝的，即使 ping 也是被拒绝的。

（1）创建虚拟机实例。

```
# nova boot --flavor 1 --image 0ee987b9-4a8e-4908-8ab5-833b781f0502 \
  --security-groups default --key-name mykey --nic net-id=NET_ID \
instance02
```

| Property | Value |
| --- | --- |
| OS-DCF:diskConfig | MANUAL |
| OS-EXT-AZ:availability_zone | nova |
| OS-EXT-STS:power_state | 1 |
| OS-EXT-STS:task_state | - |
| OS-EXT-STS:vm_state | active |
| OS-SRV-USG:launched_at | 2016-05-06T07:03:25.000000 |
| OS-SRV-USG:terminated_at | - |
| accessIPv4 | |
| accessIPv6 | |
| config_drive | |
| created | 2016-05-06T07:02:02Z |
| flavor | m1.tiny (1) |

```
| hostId                              | 70efc0aea306831e21e491e2034c43a459d57788e5c4237b2cfcfc03 |
| id                                  | e4d633ef-9102-4974-b184-15a5d76539b1                    |
| image                               | Attempt to boot from volume - no image supplied         |
| key_name                            | -                                                        |
| metadata                            | {}                                                       |
| name                                | instance02                                               |
| os-extended-volumes:volumes_attached| [{"id": "a8a1c8a1-d046-4c08-a9fa-6ad6880e0ea8"}]         |
| progress                            | 0                                                        |
| public network                      | 10.96.1.105                                              |
| security_groups                     | default                                                  |
| status                              | ACTIVE                                                   |
| tenant_id                           | 9543b7db999641fda39ff53dc22e7bcb                         |
| updated                             | 2016-05-06T09:14:26Z                                     |
| user_id                             | 33cd0cb0e80a4ba6bb5b8c36947f6267                         |
```

(2)检查虚拟机实例状态。

```
# nova list
| ID                                   | Name       | Status | Task State | Power State | Networks           |
| 09bce5e1-0cad-4b19-a678-9cfe151237ff | instance01 | ACTIVE | -          | Running     | public=10.96.1.103 |
| e4d633ef-9102-4974-b184-15a5d76539b1 | instance02 | ACTIVE | -          | Running     | public=10.96.1.105 |
```

## 2. 通过块存储设备创建虚拟机实例

通过块存储设备创建虚拟机实例可以永久保存数据，即使虚拟机实例被删除。以下是通过 nova boot 命令创建虚拟机实例的操作。

```
# nova boot --flavor FLAVOR --block-device \
source=SOURCE,id=ID,device=VOLUME_NAME,dest=DEST,size=SIZE,shutdown=PRESERVE,bootindex=INDEX
    --availability-zone=ZONE    --security-groups= SECURITY_ GROUP    --nic
net-id=NET_ID INSTANCE_NAME
```

其中，--flavor FLAVOR 参数代表主机类型的名称或 ID；source=SOURCE 参数代表创建块设备的对象来源类型，有效值包括 volume、snapshot、image 和 blank；id=ID 参数代表对象来源的 ID；device=VOLUME_NAME 参数指定需要创建的虚拟设备名称；dest=DEST 参数代表目标虚拟设备类型，有效值包括 volume 和 local；size=SIZE 参数指定创建虚拟机设备的大小；shutdown={preserve|remove}参数代表虚拟机实例被删除后卷设备的处置方式；bootindex=INDEX 参数代表卷设备的启动顺序，0 代表该块设备为第一启动设备；--availability-zone=ZONE 参数代表 OpenStack 环境中有效的 Zone；--security-groups=SECURITY_GROUP 参数代表该虚拟机实例所要使用的安全组；--nic net-id=NET_ID 参数指定虚拟机实例所要使用的网络。

(1) 创建启动镜像盘。

``` 
# cinder create --image-id $IMAGE_ID --name=VOLUME_NAME $SIZE_IN_GB
```

(2) 创建虚拟机实例。

```
# nova boot --flavor 1 \
--block-device source=volume,id=$VOLUME_ID,dest=volume,shutdown= preserve,bootindex=0 instance02
```

```
+-----------------------------------+------------------------------------------------------------+
| Property                          | Value                                                      |
+-----------------------------------+------------------------------------------------------------+
| OS-DCF:diskConfig                 | MANUAL                                                     |
| OS-EXT-AZ:availability_zone       | nova                                                       |
| OS-EXT-STS:power_state            | 1                                                          |
| OS-EXT-STS:task_state             | -                                                          |
| OS-EXT-STS:vm_state               | active                                                     |
| OS-SRV-USG:launched_at            | 2016-05-06T07:03:25.000000                                 |
| OS-SRV-USG:terminated_at          | -                                                          |
| accessIPv4                        |                                                            |
| accessIPv6                        |                                                            |
| config_drive                      |                                                            |
| created                           | 2016-05-06T07:02:02Z                                       |
| flavor                            | m1.tiny (1)                                                |
| hostId                            | 70efc0aea306831e21e491e2034c43a459d57788e5c4237b2cfcfc03   |
| id                                | e4d633ef-9102-4974-b184-15a5d76539b1                       |
| image                             | Attempt to boot from volume - no image supplied            |
| key_name                          | -                                                          |
| metadata                          | {}                                                         |
| name                              | instance02                                                 |
| os-extended-volumes:volumes_attached | [{"id": "a8a1c8a1-d046-4c08-a9fa-6ad6880e0ea8"}]        |
| progress                          | 0                                                          |
| public network                    | 10.96.1.105                                                |
| security_groups                   | default                                                    |
| status                            | ACTIVE                                                     |
| tenant_id                         | 9543b7db999641fda39ff53dc22e7bcb                           |
| updated                           | 2016-05-06T09:14:26Z                                       |
| user_id                           | 33cd0cb0e80a4ba6bb5b8c36947f6267                           |
+-----------------------------------+------------------------------------------------------------+
```

### 3. 通过ISO镜像创建虚拟机实例

OpenStack 项目支持使用 ISO 镜像创建虚拟机实例，使用 nova boot 命令启动虚拟机实例。

```
# nova boot --image ubuntu-14.04.2-server-amd64.iso \
    --block-device source=blank,dest=volume,size=10,shutdown=preserve \
    --nic net-id = NETWORK_UUID --flavor 2 INSTANCE_NAME
```

当虚拟机实例成功启动后，可以通过终端访问虚拟机实例，并根据提示安装操作系统，与在普通计算机或服务器上安装操作系统类似。当安装完成后,操作系统会重启,完成安装。在配置 IP 地址后,便可以进行远程访问。

## 26.1.2 图形界面方式

在图形界面操作中,可以根据需要从各选项中选择合适的值创建虚拟机实例。

(1) 登录 dashboard。

(2) 在 Web 界面左侧,在 Project(项目)标签上,单击 Compute(计算)标签,选中 Instance(实例)选项。

(3) 单击 Launch Instance(启动云主机)按钮。

(4) 在弹出的对话框中,选择合适的值。

Details(详情)选项如下表所示。

| 名称 | | 描述 |
|---|---|---|
| Availability Zone(可用域) | | 由云提供商设置的计算可用域 |
| Instance Name(云主机名称) | | 指定虚拟机实例名称 |
| Flavor(云主机类型) | | 指定虚拟机实例可使用的硬件资源数量 |
| Instance Count(云主机数量) | | 指定创建虚拟机实例的数量 |
| Instance Boot Source(虚拟机实例启动源) | Boot from image(从镜像启动) | 由镜像启动创建虚拟机实例 |
| | Boot from snapshot(从快照启动) | 由快照启动创建虚拟机实例 |
| | Boot from volume(从云硬盘启动) | 由块设备启动创建虚拟机实例 |
| | Boot from image (create a new volume) | 由镜像启动创建虚拟机实例,并创建一个块设备 |
| | Boot from volume snapshot (create a new volume) | 由快照启动创建虚拟机实例,并创建一个块设备 |
| Image Name(镜像名称) | | 当选择"从镜像启动"时,需指定镜像名称 |
| Instance Snapshot(云主机快照) | | 当选择"从快照启动"时,需指定云主机快照名称 |
| Volume(云硬盘) | | 当选择"从云设备启动"时,需指定云硬盘名称 |

Access&Security（访问&安全）选项如下表所示。

| 名称 | 描述 |
| --- | --- |
| Key Pair（密钥对） | 指定一个密钥对 |
| Security Group（安全组） | 指定一个安全组 |

Networking（网络）选项如下表所示。

| 名称 | 描述 |
| --- | --- |
| Selected Network（已选择的网络） | 为虚拟机实例设置网络 |

Post-Creation（创建后）选项如下表所示。

| 名称 | 描述 |
| --- | --- |
| Customization Script Source（定制脚本源） | 创建虚拟机实例后自动运行该脚本 |

Advanced Option（高级选项）选项如下表所示。

| 名称 | 描述 |
| --- | --- |
| Disk Partition（磁盘分区） | 选择手动或自动对磁盘分区 |

（5）单击 Launch（运行）按钮，虚拟机实例开始在计算节点上运行。

## 26.2 操作虚拟机实例

管理员可以管理不同租户中的虚拟机实例，查看、停止、编辑、软重启和硬重启、创建 snapshot 和迁移虚拟机实例，也可以查看虚拟机实例的日志和使用 VNC 功能本地登录虚拟机实例。

### 26.2.1 命令行方式

下面从调整虚拟机实例资源、暂停、挂起虚拟机实例、重启、删除虚拟机实例和控制台访问多个方面介绍如何管理虚拟机实例。

## 1. 调整虚拟机实例资源

(1) 使用 nova resize 命令调整虚拟机实例资源。

```
# nova resize INSTANCE_NAME FLAVOR_ID --poll
```

其中，--pool 参数显示命令执行进度。

(2) 显示虚拟机实例状态。

```
# nova list
+----------------------+------------+--------+-------------------+
| ID                   | Name       | Status | Networks          |
+----------------------+------------+--------+-------------------+
| 09bce5e1-0cad-4b1... | instance01 | RESIZE | public=10.96.1.105|
+----------------------+------------+--------+-------------------+
```

(3) 确认调整。

```
# nova resize-confirm INSTANCE_ID
```

(4) 如果虚拟机实例调整资源失败，则进行事件回滚。

```
# nova resize-revert INSTANCE_ID
```

## 2. 暂停实例

(1) 使用 nova pause 命令暂停虚拟机实例。

```
# nova pause INSTANCE_ID
```

该命令可以在内存中存储暂停实例的状态，已暂停的实例仍然可以运行。

(2) 使用 nova unpause 命令恢复被暂停的虚拟机实例。

```
# nova unpause INSTANCE_ID
```

## 3. 挂起实例

(1) 使用 nova suspend 命令挂起虚拟机实例。

```
# nova suspend INSTANCE_ID
```

(2) 使用 nova resume 命令恢复被挂起的虚拟机实例。

```
# nova resume INSTANCE_ID
```

### 4. 废弃实例

（1）使用 nova shelve 命令废弃虚拟机实例，虚拟机实例被关闭，实例本身与其相关的数据被保存，内存中的数据会丢失。

```
# nova shelve INSTANCE_ID
```

（2）使用 nova unshelve 命令恢复被废弃的虚拟机实例。

```
# nova unshelve INSTANCE_ID
```

（3）删除已废弃的虚拟机实例。

```
# nova unshelve-offload INSTANCE_ID
```

### 5. 重启实例

虚拟机实例重启有两种方式，分别是软重启和硬重启。软重启虚拟机实例是操作系统正常关闭并重启。硬重启虚拟机实例是模拟断电又加电启动的过程。

（1）使用 nova reboot 命令软重启虚拟机实例。

```
# nova reboot INSTANCE_ID
```

（2）使用 nova reboot --hard 命令硬重启虚拟机实例。

```
# nova reboot --hard INSTANCE_ID
```

（3）使用 nova rescue 命令进入救援模式。

```
# nova rescue INSTANCE_ID
```

（4）重启虚拟机实例，由救援模式进入正常模式。

```
# nova unrescue INSTANCE_ID
```

（5）虚拟机实例使用指定的镜像进入救援模式。

```
# nova rescue --rescue_image_ref IMAGE_ID SERVER
```

### 6. 删除实例

```
# nova delete INSTANCE_ID
```

### 7. 通过VNC控制台访问虚拟机实例

```
# nova get-vnc-console INSTANCE_NAME VNC_TYPE
+-------+--------------------------------------------------------------------------+
| Type  | Url                                                                      |
+-------+--------------------------------------------------------------------------+
| novnc | http://Controller:6080/vnc_auto.html?token=79e2dd6c-4474-46e7-a5aa-6354af05a5a6 |
+-------+--------------------------------------------------------------------------+
```

## 26.2.2 图形界面方式

### 1. 创建snapshot

（1）登录 dashboard。

（2）在 Web 界面左侧，在 Admin（管理员）标签上，单击 System（系统）标签，选中 Instance（实例）分类，可以看到所有的虚拟机实例。

（3）选择需创建 snapshot 的实例，在 Actions 列选择 Create Snapshot。

（4）在弹出的对话框中，填写 snapshot 名称。

（5）创建完成后，页面显示 snapshot 存放在 Image（镜像）分类中。

（6）可以选择实例从 snapshot 启动。

### 2. 控制实例的状态

（1）登录 dashboard。

（2）在 Web 界面左侧，在 Admin（管理员）标签上，单击 System（系统）标签，选中 Instance（实例）分类，可以看到所有的虚拟机实例。

（3）选择需要进行操作的实例，在 Actions 列选择需要改变的状态，如下图所示。

### 3. 查看使用概况

通过使用概况功能不仅可以查看每个虚拟机实例的资源使用情况，而且可以统计某个时间段内所有虚拟机实例的单项资源的使用情况，如 CPU、内存、磁盘容量和运行时间等。

（1）登录 dashboard。

（2）在 Web 界面左侧，在 Admin（管理员）标签上，单击 System（系统）标签，选中 Overview（概况）分类。

（3）选择某个时间段并单击 Submit（提交）按钮查看资源使用情况。

（4）单击 Download CSV Summary（下载 CSV 概要）按钮下载资源使用率表格。

## 26.3 选择主机节点运行实例

当管理员具有足够权限时，可以指定虚拟机实例运行在某个节点上，也可以指定某种角色的用户启动虚拟机实例。

（1）通过--available_zone ZONE:HOST 参数指定虚拟机实例在某个节点上运行。

```
# nova boot --image <uuid> --flavor m1.tiny --key_name test --availability-
zone nova:Compute01
```

（2）通过启用 policy.json 文件中的 create:forced_host 功能指定某种角色的用户启动虚拟机实例。默认为 admin 角色的用户启动虚拟机实例。

```
# less /etc/nova/policy.json
...
"compute:create": "",
"compute:create:attach_network": "",
"compute:create:attach_volume": "",
"compute:create:forced_host": "is_admin:True",
...
```

（3）查看有效的计算节点信息。

```
# nova hypervisor-list
+----+---------------------+-------+---------+
| ID | Hypervisor hostname | State | Status  |
+----+---------------------+-------+---------+
| 1  | Compute01           | up    | enabled |
| 2  | Compute02           | up    | enabled |
+----+---------------------+-------+---------+

# nova hypervisor-servers Compute01
+--------------------------------------+-------------------+---------------+---------------------+
| ID                                   | Name              | Hypervisor ID | Hypervisor Hostname |
+--------------------------------------+-------------------+---------------+---------------------+
| 09bce5e1-0cad-4b19-a678-9cfe151237ff | instance-00000005 | 1             | Compute01           |
+--------------------------------------+-------------------+---------------+---------------------+
```

## 26.4 计算节点配置 SSH 互信

在虚拟机实例进行资源调整或迁移时，需要配置 SSH 互信，以保证 Nova 计算服务组件使用 SSH 将数据从一个主机节点移动到其他节点。

计算节点使用同一个密钥对，实现 SSH 互信。配置步骤如下：

（1）使用第一个计算节点上 root 用户的密钥对（公钥和私钥），分别是/root/.ssh/id_rsa 和 /root/.ssh/id_ras.pub。如果没有，则可以生成密钥对。

```
# ssh-keygen -q -N ""
```

(2）修改 Nova 用户登录设置。

```
# usermod -s /bin/bash nova
```

测试 Nova 用户是否可以登录系统。

```
# su - nova
```

(3）使用 Nova 用户执行以下操作。

```
mkdir -p /var/lib/nova/.ssh
cp <PRIVATE KEY>  /var/lib/nova/.ssh/id_rsa
echo 'StrictHostKeyChecking no' >> /var/lib/nova/.ssh/config
chmod 600 /var/lib/nova/.ssh/id_rsa /var/lib/nova/.ssh/authorized_keys
```

(4）在其他计算节点上重复步骤（3）~（4）。

**注意：**
所有的计算节点必须使用相同的密钥对进行配置。

(5）在生产公钥的计算节点上执行如下命令。

```
ssh-copy-id -i <pub key> nova@remote-host
```

在 remote-host 远程计算节点上安装公钥，使用远程计算节点的 IP 地址替换 remote-host 字符串。

(6）检查 SSH 互信是否成功配置。

```
# su - nova
$ ssh *computeNodeAddress*
$ exit
```

使用其他计算节点的 IP 地址替换*computeNodeAddress*字符串。

(7）root 用户重启 Libvirt 和 nova-compute 服务。

```
# systemctl restart libvirtd.service
# systemctl restart openstack-nova-compute.service
```

## 26.5 实例热迁移

仅管理员可以操作虚拟机实例的热（在线）迁移。如果在 OpenStack 项目中设计了 cells，则同一 cell 内部可以实现该操作，cells 之间不能实现该操作。

cell 是 OpenStack 中一个非常核心的概念，主要用来解决 OpenStack 的扩展性瓶颈。众所周知，OpenStack 是由许多服务组件松耦合构成，当达到一定规模后，计算节点数量和一些服务组件（例如 DB 和 AMQP）就会有一些限制。cell 正是为解决这些限制和瓶颈而诞生的，其被设计成树形结构，所有的子 cell 公用底层 cell 的 nova-api 服务，子 cell 包含除了 nova-api 之外的所有其他 nova 服务，子 cell 和父 cell 构成完整的 nova 服务组件。每个 cell 有自己独立的 DB 和 AMQP。当然，所有的 cell 共用 keystone 服务。

实例迁移是将一个虚拟机实例从某主机节点迁移到另外一个主机节点。该功能非常强大，应用场景非常多。例如，某个主机节点需要进行维护，或需要重平衡各主机节点的压力负载等。

迁移分为冷迁移和热迁移。

(1) 冷（非在线）迁移：虚拟机实例会被关闭，然后在另一个主机节点上启动。从某种角度讲，类似于虚拟机实例执行了一次重启操作。

(2) 热（在线）迁移：几乎没有停机时间。这在某些不允许虚拟机实例停机的情况下非常适用。热迁移有几种不同方式。

- 共享存储热迁移：两种虚拟化技术都支持访问共享存储。

- 块设备热迁移：不要求共享存储，只读设备不能作为块设备进行迁移，如 CD-ROM 等。

- 卷设备热迁移：实例使用卷设备（持久性存储）而不是临时存储。不要求使用共享存储。

以下分别使用 KVM 和 XenServer 两种虚拟化技术介绍如何配置主机节点实现虚拟机实例的迁移。

## 26.5.1 KVM

### 1. 共享存储热迁移

1）准备

- 虚拟化技术：KVM-Libvirt。

- 共享存储：NOVA-INST-DIR/instances/（如/var/lib/nova/instances）目录必须能被共享存储挂载。这里以 NFS 为例，其他共享存储也可以支持，如 OpenStack 项目中的 GlusterFS 存储。

- 实例要求：被迁移的实例所使用的卷设备必须是基于 iSCSI 的。

2）备注

- 在默认情况下，Nova 计算服务不使用 Libvirt 热迁移功能。在迁移前，虚拟机实例会被暂停，停机时间大概为几分钟。

- Nova 计算服务根据磁盘迁移到内存中总数据的大小而评估时间，配置 live_migration_downtime 参数。在整个迁移过程中，时间是按照动作发生的步骤计算的，步骤之间是由数值衡量的。这就意味着步骤之间的停机时间是缩短的，但总体迁移的时间是增加的。相对而言，成功完成迁移所消耗的时间缩短了。

- 本章假设 instances_path 值在/etc/nova/nova.conf 文件中配置。如果 state_path 或 instances_path 值被修改，则相应修改其他相关命令。

- 必须指定 vncserver_listen=0.0.0.0，否则迁移不能正常运行。

- 必须在每个主机节点上指定运行 nova-compute 服务的 instances_path 值，instances_path 的挂载点必须在每个计算节点都是相同的，否则迁移不能正常运行。

3）Nova 计算服务环境介绍：

- 三个主机节点，分别为 Controller、Compute01 和 Compute02。Controller 是控制节点，其上运行 nova-api、nova-scheduler、nova-network、cinder-volume 和 nova-objectstore。Compute01 和 Compute02 是计算节点，其上分别运行 nova-compute 和 cinder-volume。务必确认 nova.conf 文件中的 state_path 值在每个主机节点都是相同的。

- 本例中，Controller 是 NFSv4 服务器端，导出 NOVA-INST-DIR/instances 目录；Compute01 和 Compute02 是 NFSv4 客户端，挂载 Controller 的导出目录。

4）配置

（1）配置每个主机节点的 DNS 或/etc/hosts 文件，以便可以在每个主机节点上进行名字解析。可以使用 ping 命令在每个主机节点上进行测试。

```
# ping Controller
# ping Compute01
# ping Compute02
```

（2）确保每个主机节点的 Nova 计算服务和 Libvirt 用户的 UID 和 GID 是相同的，从而确保 NFS 挂载权限的正确性。

（3）确保 Compute01 和 Compute02 彼此之间 Nova 用户使用 SSH 访问时不需要密码，建立信任。因为计算节点之间在迁移时不仅需要复制文件，而且需要侦测源主机节点和目标主机节点是否挂载共享存储。

（4）导出 Controller 节点上的 NOVA-INST-DIR/instances（/var/lib/nova/instance）目录，确保主机节点 Compute01 和 Compute02 上的 Nova 用户对其可读可写。

（5）在 Controller 节点上配置 NFS 共享，添加以下内容到/etc/exports 文件中。

```
NOVA-INST-DIR/instances    IP/NETMASK(rw,sync,fsid=0,no_root_squash)
```

使用合适的信息替换 IP/NETMASK 字符串。根据本书的网络设置，可以替换成 172.168.1.0/255.255.255.0。

（6）重启 NFS 服务。

```
# /etc/init.d/nfs-kernel-server restart
# /etc/init.d/idmapd restart
```

（7）在所有主机节点上修改 NOVA-INST-DIR/instances 目录的权限。

```
# chmod o+x NOVA-INST-DIR/instances
```

（8）编辑 Compute01 和 Compute02 主机节点的/etc/fstab 文件，添加以下内容。

```
Controller:/NOVA-INST-DIR/instances    /NOVA-INST-DIR/instances nfs4
defaults 0 0
```

(9）检查设置是否正确。

```
# mount -a -v
```

(10）在所有主机节点上检查 NOVA-INST-DIR/instances/ 目录的权限设置。

```
# ls -ld NOVA-INST-DIR/instances/
drwxr-xr-x 2 nova nova 4096 2016-05-19 17:21 nova-install-dir/instances/
```

(11）更新 Libvirt 关于安全的配置，如下（不详细展开）：

① SSH tunnel to libvirtd's UNIX socket（连接 Libvirtd UNIX Socket 的 SSH 通道加密）。

② libvirtd TCP socket, with GSSAPI/Kerberos for auth+data encryption（利用 GSSAPI/Kerberos 进行身份认证和数据加密）。

③ libvirtd TCP socket, with TLS for encryption and X509 client certs for authentication（利用 TLS 进行加密和 X509 客户端认证）。

④ libvirtd TCP socket, with TLS for encryption and Kerberos for authentication（利用 TLS 进行加密和 Kerberos 认证）。

⑤ 重启 Libvirt。

```
# stop libvirt-bin && start libvirt-bin
```

(12）在所有主机节点上为 Libvirt 配置防火墙通信。Libvirt 监听 TCP 端口 16509，KVM 通信使用 TCP 端口范围为 49152～49261。

(13）编辑 nova.conf 文件，设置迁移操作所需的停机时间。

```
live_migration_downtime = 500
live_migration_downtime_steps = 10
live_migration_downtime_delay = 75
```

- live_migration_downtime 参数用于设置允许的最长停机时间，单位是毫秒，默认值是 500。

- live_migration_downtime_steps 参数用于设置迁移完成所需的步骤总数，默认值是 10。

- live_migration_downtime_delay 参数用于设置每个步骤之间等待的最长时间，单位是秒，默认值是 75。

5）实现热迁移

➢ 在 Liberty 版本中，Nova 计算服务默认不使用 Libvirt 热迁移功能，需要在 nova.conf 文件的[libvirt]项取消注释。

```
live_migration_flag=VIR_MIGRATE_UNDEFINE_SOURCE,VIR_MIGRATE_PEER2PEER,
VIR_MIGRATE_LIVE,VIR_MIGRATE_TUNNELLED
```

在 Kilo 版本以前，Nova 计算服务默认不使用 Libvirt 热迁移功能，因为其中存在一个风险：热迁移操作一直进行，无法完成。该风险发生的原因是数据写入磁盘的速度比数据迁移的速度快。

## 2．块设备热迁移

块设备热迁移与共享存储热迁移基本相同，唯一不同的是块设备热迁移目录 NOVA-INST-DIR/instances 在本地而不是共享的，也就没有 NFS 服务器端和客户端。

> 注意：
> - 在进行块设备热迁移时，必须使用--block-migrate 参数。
> - 只读设备不能作为块设备进行热迁移，如 CD-ROM。
> - 因为临时驱动器是通过网络进行数据复制的，所以当数据写入速度比数据迁移速度快时，热迁移会一直进行，无法完成。

## 3．卷设备热迁移

虚拟机实例使用卷设备进行热迁移和共享存储类似，迁移简单且速度快。下面是卷设备热迁移的配置步骤。

（1）创建基于 Image 的卷设备。

```
# cinder create --image-id 0ee987b9-4a8e-4908-8ab5-833b781f0502 --display-name boot_volume01_cirros
```

（2）创建虚拟机实例。

```
# nova boot --flavor m1.tiny --block_device_mapping vda=a8a1c8a1-d046-4c08-a9fa-6ad6880e0ea8:::0 instance02
```

（3）检查创建的虚拟机实例状态。

```
# nova show e4d633ef-9102-4974-b184-15a5d76539b1
```

```
+-----------------------------------+------------------------------------------------------+
| Property                          | Value                                                |
+-----------------------------------+------------------------------------------------------+
| OS-DCF:diskConfig                 | MANUAL                                               |
| OS-EXT-AZ:availability_zone       | nova                                                 |
| OS-EXT-STS:power_state            | 1                                                    |
| OS-EXT-STS:task_state             | -                                                    |
| OS-EXT-STS:vm_state               | active                                               |
| OS-SRV-USG:launched_at            | 2016-05-06T07:03:25.000000                           |
| OS-SRV-USG:terminated_at          | -                                                    |
| accessIPv4                        |                                                      |
| accessIPv6                        |                                                      |
| config_drive                      |                                                      |
| created                           | 2016-05-06T07:02:02Z                                 |
| flavor                            | m1.tiny (1)                                          |
| hostId                            | 70efc0aea306831e21e491e2034c43a459d57788e5c4237b2cfcfc03 |
| id                                | e4d633ef-9102-4974-b184-15a5d76539b1                 |
| image                             | Attempt to boot from volume - no image supplied      |
| key_name                          | -                                                    |
| metadata                          | {}                                                   |
| name                              | instance02                                           |
| os-extended-volumes:volumes_attached | [{"id": "a8a1c8a1-d046-4c08-a9fa-6ad6880e0ea8"}]  |
| progress                          | 0                                                    |
| public network                    | 10.96.1.105                                          |
| security_groups                   | default                                              |
| status                            | ACTIVE                                               |
| tenant_id                         | 9543b7db999641fda39ff53dc22e7bcb                     |
| updated                           | 2016-05-06T09:14:26Z                                 |
| user_id                           | 33cd0cb0e80a4ba6bb5b8c36947f6267                     |
+-----------------------------------+------------------------------------------------------+
```

（4）进行 instance02 实例热迁移。

```
# nova live-migration instance02
```

 错误：

在进行基于卷设备的虚拟机实例热迁移时，出现以下错误。

```
    May 05 20:47:24 Compute02 libvirtd[11812]: internal error: QEMU / QMP
failed: Could not access KVM kernel module: No such file or directoryfailed
to initialize KVM: No such file or directory
    May 05 20:47:24 Compute02 libvirtd[11812]: Failed to probe capabilities
for /usr/bin/qemu-kvm: internal error: QEMU / QMP failed: Could not access
KVM kernel module: No such file or directory failed to initialize KVM: No
such file or directory
    May 05 20:47:25 Compute02 libvirtd[11812]: Failed to probe capabilities
for /usr/bin/qemu-kvm: internal error: QEMU / QMP failed: Could not access
KVM kernel module: No such file or directoryfailed to initialize KVM: No such
file or directory
    May 05 20:47:42 Compute02 libvirtd[11812]: Failed to probe capabilities
for /usr/bin/qemu-kvm: internal error: QEMU / QMP failed: Could not access
KVM kernel module: No such file or directory failed to initialize KVM: No
such file or directory
    May 05 20:47:42 Compute02 libvirtd[11812]: internal error: QEMU / QMP
```

```
failed: Could not access KVM kernel module: No such file or directory failed
to initialize KVM: No such file or directory
    May 05 20:47:42 Compute02 libvirtd[11812]: Failed to probe capabilities
for /usr/bin/qemu-kvm: internal error: QEMU / QMP failed: Could not access
KVM kernel module: No such file or directory failed to initialize KVM: No
such file or directory
    May 06 15:28:08 Compute02 libvirtd[11812]: operation failed: Failed to
connect to remote libvirt URI qemu+tcp://Compute01/system: unable to connect
to server at 'Compute01:16509': Connection refused
    May 06 15:30:03 Compute02 libvirtd[11812]: operation failed: Failed to
connect to remote libvirt URI qemu+tcp://Compute01/system: unable to connect
to server at 'Compute01:16509': Connection refused
    May 06 15:38:00 Compute02 libvirtd[11812]: operation failed: Failed to
connect to remote libvirt URI qemu+tcp://Compute01/system: unable to connect
to server at 'Compute01:16509': Connection refused
```

基于以上错误信息推测,可能存在两种原因。

(1)防火墙问题,拒绝对端口的连接。

(2)配置不正确,该端口没有启动。

经检查,属于第二种原因,那么需要在计算节点上进行如下配置。

(1)编辑/etc/libvirt/libvirtd.conf 文件,添加如下内容。

```
listen_tls = 0
listen_tcp = 1
auth_tcp = "none"
```

(2)编辑/etc/sysconfig/libvirtd 文件,添加如下内容。

```
LIBVIRTD_ARGS="-listen"
```

(3)重启 Libvirtd。

```
systemctl restart libvirtd.service
```

(4)编辑/etc/nova/nova.conf 文件,取消如下内容的注释。

```
live_migration_flag=VIR_MIGRATE_UNDEFINE_SOURCE,VIR_MIGRATE_PEER2PEER,VIR_MIGRATE_LIVE,VIR_MIGRATE_TUNNELLED
```

重新尝试后,操作成功。

## 26.5.2 XenServer

### 1. 共享存储热迁移

1）准备

- 共享存储：NFS 服务器共享文件系统，所有 NFS 客户端均可见。
- NFS 支持版本：NFS v3 over TCP/IP。

2）配置

（1）主 XenServer 添加 NFS VHD 存储，并设置其为默认的存储资源池。

（2）编辑 nova.conf 文件，添加以下内容，使所有的计算节点都使用默认的存储资源池。

```
sr_matching_filter=default-sr:true
```

（3）创建主机节点聚合。

```
# nova aggregate-create POOL_NAME AVAILABILITY_ZONE
```

（4）主机节点聚合添加 metadata，标记为虚拟机管理池。

```
# nova aggregate-set-metadata AGGREGATE_ID hypervisor_pool=true
# nova aggregate-set-metadata AGGREGATE_ID operational_state=created
```

（5）添加第一个计算节点并成为该聚合的一部分。

```
# nova aggregate-add-host AGGREGATE_ID MASTER_COMPUTE_NAME
```

（6）添加其他计算节点到该聚合中。

```
# nova aggregate-add-host AGGREGATE_ID COMPUTE_HOST_NAME
```

注意：
　　计算节点将会被关闭以加入 XenServer 池中。如果有虚拟机实例运行在计算节点上，则该操作会失败。

### 2. 块存储热迁移

XenServer 虚拟化管理程序：虚拟化管理程序必须支持存储 XenMotion 功能。

> **注意**:
> 块设备热迁移和本地存储池一起协同运行，虚拟机实例不能挂载其他卷设备。

## 26.6 实例冷迁移

虚拟机实例可以在计算节点之间迁移，以平衡计算资源的使用。nova-scheduler 服务可以为虚拟机实例选择迁移到的目标主机节点。该迁移功能不关注虚拟机实例和目标主机节点是否共享存储。计算节点之间必须建立 SSH 互信，以保证磁盘数据成功迁移。

（1）选择需要迁移的虚拟机实例。

```
# nova list
```

（2）查看需要迁移的虚拟机实例信息。

```
# nova show VM_ID
```

（3）迁移虚拟机实例。

```
# nova migrate VM_ID
```

（4）以下脚本可以观察到虚拟机实例的实时迁移过程。

```
#!/bin/bash

# Provide usage
usage() {
echo "Usage: $0 VM_ID"
exit 1
}

[[ $# -eq 0 ]] && usage

# Migrate the VM to an alternate hypervisor
echo -n "Migrating instance to alternate host"
VM_ID=$1
nova migrate $VM_ID
```

```
VM_OUTPUT=`nova show $VM_ID`
VM_STATUS=`echo "$VM_OUTPUT" | grep status | awk '{print $4}'`
while [[ "$VM_STATUS" != "VERIFY_RESIZE" ]]; do
echo -n "."
sleep 2
VM_OUTPUT=`nova show $VM_ID`
VM_STATUS=`echo "$VM_OUTPUT" | grep status | awk '{print $4}'`
done
nova resize-confirm $VM_ID
echo " instance migrated and resized."
echo;

# Show the details for the VM
echo "Updated instance details:"
nova show $VM_ID

# Pause to allow users to examine VM details
read -p "Pausing, press <enter> to exit."
```

> **错误:**
>
> 在进行虚拟机实例迁移时，出现以下错误。
>
> ERROR (Forbidden): Policy doesn't allow compute_extension:admin_actions: migrate to be performed. (HTTP 403)
>
> 原因是操作实例迁移的用户权限不正确，必须是管理员。
>
> 解决方法：修改 policy.json 文件，允许某用户迁移实例。
>
> ProcessExecutionError: Unexpected error while running command.
> Stderr: u Host key verification failed.\r\n
>
> 原因是计算节点之间的 SSH 互信设置不正确。

## 26.7 实例转移

当主机节点硬件故障或者出现错误而导致计算节点宕机，不能提供服务时，可以通过转移虚拟机实例使其可用。可以手动指定转移虚拟机实例到某个目标主机节点上。如果不指定，则 nova-scheduler 服务会自动选择目标主机节点运行该虚拟机实例。

为确保用户生产数据的安全，需要在目标主机节点上配置共享存储。在转移虚拟机实例之前，Nova 计算服务会检查共享存储是否在目标主机节点可见并有效。在进行转移虚拟机实例操作时，一定要确认虚拟机实例所在的主机节点不可操作，各项服务均失败，否则转移操作会失败。

（1）列表显示所有的主机节点。

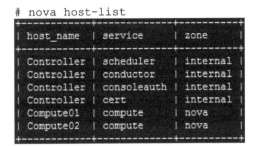

（2）转移虚拟机实例。可以指定--password PWD 选项通过实例的密码验证，当然，也可以不指定该参数。成功转移实例后，命令会自动生成一串密码。以下操作将实例转移到 HOST_B 节点上。

该命令会保留原始配置并返回一串密码。原始配置包括实例 ID、名称、UID 和 IP 地址等。

（3）为保存失败主机节点上虚拟机实例中的用户数据，需要配置共享存储，命令如下。

```
# nova evacuate EVACUATED_SERVER_NAME HOST_B --on-shared-storage
```

> **注意：**
>
> 本章介绍到此，关于虚拟机实例整体的迁移操作分为三类，彼此之间既有相似又有区别，下面详细描述三种迁移类型的区别。
>
> **1. 实例热迁移**
>
> 特点：实例在迁移过程中可以对外提供访问，可以持续读/写磁盘数据，网络正常连接，即使出现闪断，时间应该是毫秒级的，非常短暂。
>
> 应用场景：业务系统不允许停机；重平衡主机节点的计算资源；主机节点进行停机维护。
>
> 操作命令：
>
> ```
> nova live-migrate
> ```
>
> **2. 实例冷迁移**
>
> 特点：实例在迁移过程中网络中断，磁盘暂时不能读/写数据，系统停止使用，迁移完成后自动恢复。
>
> 应用场景：重平衡主机节点的计算资源；主机节点进行停机维护。
>
> 操作命令：
>
> ```
> nova migrate
> ```
>
> **3. 实例转移**
>
> 特点：实例转移发生在计算节点宕机的情况下，实例可以转移到可用的目标计算节点上。当然，在此种情况下，磁盘、网络等均不可用。
>
> 应用场景：计算节点宕机；网卡故障；主机节点的Nova计算服务不可用。
>
> 操作命令：
>
> ```
> nova evacuate
> ```

# 第 27 章 OpenStack 版本升级

众所周知，OpenStack 版本每 6 个月更新一次，目前的最新版本是 Mitaka。版本更新频繁，功能特性增多，随之带来的 OpenStack 版本升级工作难度大、危险高。本章从多个维度力求提供一套完整的升级方案，最大限度地保证升级工作的顺利进行、基础架构的安全和稳定。

## 27.1 升级准备

本章从升级计划、升级测试和升级级别三个方面阐述升级前的准备。

### 1. 升级计划

- 查看 http://releases.openstack.org/ 网站，了解 OpenStack 各版本的发布信息，功能特性的升级、更新和弃用等，重点关注不同版本、相同服务组件的兼容性。
- 评估升级操作对用户的影响。升级操作会使整个 OpenStack 项目的管理失灵，一些操作不能正常完成。根据经验发现，正确而全面的准备工作会使正在升级的 OpenStack 项目依然可用，但是不可避免地会出现虚拟机实例、网络、存储等资源瞬间中断。

- 评估升级操作的正确方式。升级操作的方式有很多种，例如，可以从虚拟机实例的维度进行划分，利用热迁移，将某计算节点上的所有虚拟机实例临时迁移到其他计算节点上。但这是一种很危险的方式，因为不能确定升级过程中数据库的持续可用性，这样 OpenStack 环境会变得非常不稳定。在进行升级操作前，必须通知用户对与其相关的数据进行备份，这点非常重要。
- 评估正在采用的基础架构的合理性，看看是否需要更新或合并某些配置。
- 分析可能升级失败的场景和原因，准备方案进行应对，使其能够完整、安全地回滚到升级前的版本，包括数据库、配置文件和软件包。
- 准备一套类似生成环境的 OpenStack 测试环境，对其进行版本升级，总结出一套完整的、切实可行的方式方法。

### 2. 升级测试

版本升级前的完整测试对后续生产环境的升级至关重要。升级的版本选择也非常重要，当某一版本发布后，立即进行升级，一些未知的 Bug 会极大地阻碍升级进度。根据经验发现，升级到次新版本是比较安全的选择。

在升级测试中，即使多套 OpenStack 环境都是严格依据官方文档进行配置的，但是彼此之间仍有不同，仍然需要对这些版本的 OpenStack 进行全面测试。

综上所述，进行完整而全面的测试，并不是说使用与生产环境完全相同的硬件资源，而是需要从架构角度整体考虑。以下两点会最大限度地降低升级测试的成本。

- 使用私有云：最简单的方式就是使用硬件设备搭建一套全新的 OpenStack 环境，但是这样看起来有点资源浪费，但是对于版本升级非常快速、有效。
- 使用公有云：大部分公有云按照时间收费，可以使用公有云资源测试 OpenStack 环境中控制节点的资源扩展限制，从而有效地降低成本和缩短时间。

关于测试环境的搭建，有以下两种方式可以借鉴。

- 依据 http://docs.openstack.org/index.html#install-guides 网站的操作手册，完全手动安装各个节点。

- 使用克隆技术自动创建节点，按照配置修改相关文件，使其正常工作。

这两种方式都是有效的，管理员可以根据自己的经验和需求进行选择。OpenStack 测试环境的升级是否成功有很多影响，对 OpenStack 环境的使用方式和用户对其的交互方式都会产生影响。

以上都是针对软件包的升级，对于数据库中数据的测试也比较重要。一些 MySQL 的 Bug 在软件升级过程中可能不会发现，因为全新安装的数据库和升级后数据库中的数据表会有些许不同，对长期运行、数据量较大的数据库会产生一些影响，这对 OpenStack 生产环境的版本升级影响巨大。

依靠人工进行的测试项和注意项大致如上。OpenStack 环境版本升级成功后，需重点关注其性能，毕竟一次成功的版本升级不是只以功能正确为目的的。

### 3. 升级级别

从 OpenStack 的 G（Grizzly）版本开始，一个新特性"升级级别（Upgrade Level）"被加入到 OpenStack 的 Nova 计算服务组件中，它能够在不同的 Nova 计算服务组件版本之间进行 RPC（消息队列）通信。

由于当前 OpenStack 中对象化的引入，Nova 计算服务中的 API、Scheduler、Conductor 和 Compute 进程之间存在相互依赖，无法简单描绘出一张依赖关系图。而目前 OpenStack 版本每 6 个月更新一次，版本升级又是一个无法绕过的命题，所以进程间的循环依赖给版本升级造成很大的障碍。升级级别功能就是为了解决这个难题而生的，当进行在线升级时，它允许不同版本的 Nova 正常通信、正确运行。

例如，在没有升级级别功能时，X+1 版本的 Nova 计算服务组件能够接收和翻译 X 版本的 RPC 消息，但是它仅能发送 X+1 版本的 RPC 消息，那么 X 版本的 Nova 计算服务组件就不能翻译 X+1 版本的 RPC 消息。

在进行版本升级时，管理员可以在 nova.conf 文件中进行配置，以使不同版本的 RPC 消息兼容，使在线升级顺利进行。配置文件 nova.conf 中允许指定 RPC 版本号，支持版本别名设置。例如：

```
[upgrade_levels]
```

```
compute=X+1
conductor=X+1
scheduler=X+1
```

以上设置允许在升级过程中的 X 版本与 X+1 版本的 RPC 正常通信。一旦升级完成，管理员可以删除以上配置选项。

## 27.2 版本升级

本节具体阐述版本升级的操作步骤。其中，Neutron 网络服务组件的升级有一定的特殊性，因而首先介绍一下 Neuron 的升级策略。

### 1. Neutron升级策略

Neutron 网络服务组件的升级支持两种方式。

- 停止所有服务，然后升级代码，最后启动服务。
- 根据维护窗口，循序渐进，逐渐升级所有服务。

推荐采用第二种升级方式，它允许将升级分割成多个任务，从而带来更少的停机时间。通常将第二种升级方式称为"滚动升级"。

滚动升级方式将版本升级工作分配在多个时间段内进行，造成在同一个 OpenStack 项目中同时存在多个代码版本，而且彼此之间需要交互和通信，从而给升级工作带来一些约束。

- 新旧版本间必须可以正常交互和通信。
- 旧版本数据库的架构不能严格，必须兼容新版本。

以上约束可以通过以下两种方式解决。

- neutron-server 组件必须具有向下或向后的兼容性。
- 隔离独立的服务。

根据以上说明和解释，具体的操作顺序就是：首先升级所有的 neutron-server；其次升级所有的 neutron-agent。

根据以上确定的升级顺序，下面开始介绍 neutron-server 的升级方法。

（1）neutron-server 升级。

neutron-server 是第一个更新到新版本的组件，也是唯一一个依赖于数据库的组件，其他组件都通过 AMQP 进行通信，并且不依赖于数据库。

首先进行数据库升级，分为两部分：

- neutron-db-manage upgrade –expand。
- neutron-db-manage upgrade –contract。

第一步可以在 neutron-server 运行时执行，第二步需要在所有的 neutron-server 关闭时执行。完成后，启动所有的 neutron-server。

（2）neutron-agent 升级。

如果在 OpenStack 项目中没有使用 AMQP，那么这部分可以直接略过。其他关于升级的章节内容依然适用。

当 neutron-server 升级完成后，开始升级 neutron-agent。此时，新版本的 neutron-server 通过 AMQP 正在与旧版本的 neutron-agent 进行交互和通信。

neutron-agent 的推荐升级顺序是：

- L2 agent 升级（openvswitch、linuxbridge、sr-iov）。
- 其他 agent 升级（L3、DHCP、metadata……）。

计算节点和网络节点都有各自的升级顺序，与其他节点之间没有任何依赖关系。

## 2. 准备

（1）检查 OpenStack 项目的状态，看是否有一些错误。例如，是否有虚拟机实例在执行删除操作后没有被完全丢弃，资源没有被释放。

(2)确认 Neutron 网络服务组件使用的数据库版本。

```
# su -s /bin/sh -c "neutron-db-manage --config-file /etc/neutron/ neutron.conf \
    --config-file /etc/neutron/plugins/ml2/ml2_conf.ini current" neutron
```

## 3. 执行备份

(1)备份所有主机节点上的配置文件,根据实际情况可以灵活调整。例如:

```
# for i in keystone glance nova neutron openstack-dashboard cinder heat ceilometer; \
    do mkdir $i-kilo; \
    done
# for i in keystone glance nova neutron openstack-dashboard cinder heat ceilometer; \
    do cp -r /etc/$i/* $i-kilo/; \
    done
```

(2)对数据库中的数据执行全备份。例如:

```
# mysqldump -u root -p --opt --add-drop-database --all-databases > juno-db-backup.sql
```

## 4. 配置更新源

以下操作需要在所有节点上运行:

- 删除节点上的旧源。
- 在节点上添加新源。
- 更新源数据,使其生效。

## 5. 升级软件包

根据实际情况和具体配置,需要在每个主机节点上升级相关软件包。软件包升级后可能需要重启,从而打断该软件包提供的服务。例如,使用 TGT iSCSI 作为 Cinder 块存储连接技术,在更新 iSCSI 软件包重启该服务时,可能会造成虚拟机实例上该卷设备不可用。

操作系统的软件包管理器会在某个软件包更新后，在与该软件包相关的旧配置文件名后添加后缀，进行标注，方便后续对这些文件的查看和调整。

## 6. 更新OpenStack服务组件

在主机节点上更新 OpenStack 的相关服务组件时，一般需要修改多个配置文件。每个服务组件进行更新时不尽相同，可能有多个步骤，如停止服务、同步数据库 Schema、启动服务等。强烈建议在执行更新操作时，务必在确认本服务组件正确更新后再进行下一个服务组件的更新。

服务组件的更新需要按照一定的顺序进行，具体如下。

控制节点：

- Keystone 身份认证服务组件。
- Glance 镜像服务组件。
- Nova 计算服务组件（如果使用 nova-network 服务组件，则也包括在其中）。
- Neutron 网络服务组件。
- Cinder 块存储服务组件。
- Horizon 图形界面服务组件（在标准配置下，更新 Horizon 图形界面服务组件后重启 Apache HTTP 服务立即生效）。
- Heat 模板服务组件。
- Ceilometer 计量服务组件（在标准配置下，更新 Ceilometer 计量服务组件后重启服务立即生效，不需要修改其他配置文件）。
- 修改 Nova 计算服务组件配置文件，重启服务生效。
- 修改 Neutron 网络服务组件配置文件，重启服务生效。

块存储节点：

Cinder 块存储服务组件：更新 Cinder 块存储服务组件，重启服务生效。

计算节点：

Neutron 网络服务组件：修改配置文件，重启服务生效。

### 7．最后步骤

完成以上所有的升级步骤后，需要执行以下操作。

- 在计算节点上修改/etc/nova/nova.conf 配置文件，设置 DHCP 为执行升级操作前的值。
- 更新所有的.ini 配置文件的密码和管道设置，以匹配自己的 OpenStack 环境。
- 升级完成后，用户通过 nova image-list 和 glance image-list 命令看到的结果与升级前可能不同。如果不同，则需要检查/etc/glance/policy.json 和/etc/nova/policy.json 文件是否包含"context_is_admin": "role:admin"选项，该选项可以控制对镜像的访问权限。
- 在 OpenStack 环境中进行测试，同时通知用户，检查运行是否正常。

## 27.3 版本回退

OpenStack 项目的版本升级涉及一系列复杂的操作，存在失败的可能性。这就引出另外一个命题：如何在升级失败后，快速而完整地恢复？例如，在升级前，需要对 OpenStack 生成环境的数据库数据执行全备份。

本节对 OpenStack 版本的回退提供一些指导和操作步骤。每个版本基本相似，可以共同参考本节内容。

版本回退现象出现的一个经典场景就是在升级 OpenStack 生成环境版本时遇到了多个错误，而在之前的 OpenStack 测试环境升级时没有出现，多次尝试修复无果后，可能需要进行 OpenStack 版本的回退操作。在此场景下，大概需要三个步骤完成版本的回退。

（1）修改配置文件，回退到升级之前的内容。

（2）恢复数据库。

（3）降级所有的软件包。

再次重申：版本升级前务必要进行数据备份，版本回退前一定要确认备份数据存在并完整。版本回退是一个非常艰难的过程，可能需要付出比版本升级更多的努力和时间去解决版本回退中出现的问题。在做出充分的权衡和评估后，版本回退是最后的选择。

下面介绍回退操作的步骤。

（1）停止所有的 OpenStack 服务组件。

（2）将备份的关于各服务组件的目录和文件恢复到原有目录（/etc/<service_name>）中。

（3）使用之前的数据库备份（kilo-db-backup.sql）恢复数据库数据。

```
# mysql -u root -p < RELEASE_NAME-db-backup.sql
```

（4）降级 OpenStack 软件包版本。

 警告：

降级 OpenStack 软件包版本是整个版本回退操作中最危险的步骤，能否成功与 OpenStack 版本和操作系统整体管理、配置息息相关。

① 首先确认操作系统中已经安装了全部 OpenStack 软件包。可以使用 rpm -qa 命令搜索所有已安装的软件包，并输出到一个标准文件中。例如，可以使用以下命令在控制节点上搜索所有已安装的软件包。

```
rpm -qa|grep -E "neutron|keystone|glance|nova|cinder" >> /tmp/openstack_package_installed_list
    python-neutron-7.0.4~a0~dev19-1.1.noarch
    python-cinderclient-1.4.0-2.2.noarch
    openstack-nova-consoleauth-12.0.2~a0~dev20-1.1.noarch
    openstack-cinder-scheduler-7.0.2~a0~dev30-2.1.noarch
    python-keystoneclient-1.7.2-6.2.noarch
    openstack-keystone-8.0.2~a0~dev34-1.3.noarch
    python-glanceclient-1.1.0-2.2.noarch
    python-nova-12.0.2~a0~dev20-1.1.noarch
    openstack-neutron-metadata-agent-7.0.4~a0~dev19-1.1.noarch
    python-neutronclient-3.1.1-1.2.noarch
    openstack-nova-conductor-12.0.2~a0~dev20-1.1.noarch
    openstack-neutron-linuxbridge-agent-7.0.4~a0~dev19-1.1.noarch
```

```
openstack-cinder-7.0.2~a0~dev30-2.1.noarch
python-keystonemiddleware-2.3.1-1.2.noarch
python-glance_store-0.9.1-2.2.noarch
python-glance-11.0.2~a0~dev7-1.1.noarch
openstack-nova-12.0.2~a0~dev20-1.1.noarch
openstack-neutron-server-7.0.4~a0~dev19-1.1.noarch
openstack-neutron-7.0.4~a0~dev19-1.1.noarch
openstack-neutron-dhcp-agent-7.0.4~a0~dev19-1.1.noarch
openstack-cinder-api-7.0.2~a0~dev30-2.1.noarch
python-keystone-8.0.2~a0~dev34-1.3.noarch
openstack-nova-scheduler-12.0.2~a0~dev20-1.1.noarch
python-cinder-7.0.2~a0~dev30-2.1.noarch
openstack-glance-11.0.2~a0~dev7-1.1.noarch
openstack-nova-api-12.0.2~a0~dev20-1.1.noarch
python-novaclient-2.30.1-7.1.noarch
openstack-nova-novncproxy-12.0.2~a0~dev20-1.1.noarch
python-keystoneauth1-1.1.0-1.5.noarch
openstack-nova-cert-12.0.2~a0~dev20-1.1.noarch
```

根据实际配置和不同功能的主机节点设置，以上输出内容可能不尽相同。

② 配置软件包仓库，确认旧版的 OpenStack 软件包版本。

```
for package in `cat /tmp/openstack_package_installed_list`
do
    rpm -qi $package|grep Name|awk '{print $3}' >> /tmp/openstack_package_installed_name_list
done

for package_name in `cat /tmp/openstack_package_installed_name_lis`
do
    name=$package_name
    ver_info=`zypper info $package_name|grep Status|awk '{print $4}'`
    package_full_name=$name-${ver_info}.noarch
    echo $package_full_name >> /tmp/openstack_package_zypper_list
done
```

以上脚本会检查操作系统当前 OpenStack 版本和旧版 OpenStack 版本信息，供管理员进行比对，以确认已经升级和未升级的软件包信息，并将已经升级的软件包信息输入到指定文件中。

```
for compare_name in `cat /tmp/openstack_package_zypper_list`
do
    cat /tmp/openstack_package_installed_list|grep $compare_name
    if [ $? != 0 ]
        then
            echo $compare_name >> /tmp/compare_list
        fi
done
```

③ 安装旧版本的 OpenStack 软件包。

```
zypper install -f </tmp/compare_list>
```

至此，整个回退操作全部完成，但是 OpenStack 环境的正常运行可能还需要完成后续的许多工作。

# 第 28 章 故障排查

## 28.1 计算服务组件故障排查

以下针对 OpenStack 环境中的一些常见错误提供解决方案。当然，也可以访问 https://ask.openstack.org 网址提问或查找问题答案。网站中有其他用户提出的问题，许多 Bug 也被修复，一些资源能够帮助用户更好地解决问题。

**问题 1**：确实凭据产生 403 forbidden 错误。

**解决方案**：

- 手动方法：从租户 ZIP 文件中取得 novarc，备份已经存在的 novarc 文件，手动导入该 novarc 文件。

- 脚本方法：从租户 ZIP 文件中取得 novarc，并导入该 novarc 文件。

在首次运行 nova-api 服务时，会生产认证信息，包括 openssl.cnf。在运行 nova-api 服务之前启动 CA 服务，ZIP 文件不能被生成。重启服务，当 CA 服务变成可用状态时，创建 ZIP 文件。

**问题 2**：虚拟机实例状态为 pending，或无法通过 SSH 网络连接到虚拟机实例。

**解决方案**：替换镜像重新生成虚拟机实例。如 Ubuntu，它支持在 FlatManager 网络模式下成功获取 IP 地址，但有的镜像会存在问题。

查看虚拟机实例是否始终处于 spawning 等非正常状态。在 nova-compute 主机节点上查看/var/lib/nova/instances 目录下的 libvirt.xml、disk、disk-raw、kernel、ramdisk 和 console.log（虚拟机实例启动后才会生成该文件）文件是否存在或大小是否正常。

任何文件为空或大小不正确，虚拟机实例都会存在问题。

检查虚拟机实例的日志文件，包括/var/log/libvirt/qemu、/var/log/nova 等。

当执行以下命令时，是否会返回一个错误。

```
# virsh create libvirt.xml
```

**问题 3**：虚拟机实例正在运行，但是无论是在 dashboard 还是在命令行中都不产生日志输出，或者出现乱码。

**解决方案**：添加以下内容到虚拟机实例 bootloader 的内核参数中。

```
console=tty0 console=ttyS0,115200n8
```

重启虚拟机实例，日志输出正常。

**问题 4**：虚拟机实例始终保持一个中间状态，如 deleting、spawning 等。

**解决方案**：使用 nova reset-state 命令重置虚拟机实例的状态。

```
# nova reset-state c6bbbf26-b40a-47e7-8d5c-eb17bf65c485
# nova delete c6bbbf26-b40a-47e7-8d5c-eb17bf65c485
```

或者重新激活虚拟机实例，使其状态恢复正常。

```
# nova reset-state --active c6bbbf26-b40a-47e7-8d5c-eb17bf65c485
```

**问题 5**：管理员使用 libverity（版本 1.2.2）创建实时快照报错。

**解决方案**：在 nova.conf 文件的[workarounds]项下，设置 disable_libvirt_livesnapshot 值为 True，关闭实时快照。

```
[workarounds]
disable_libvirt_livesnapshot = True
```

## 28.2 块存储服务组件故障排查

大多数块存储错误都是由不正确的配置导致的,为了解决这些问题,可以查看日志文件 cinder-api(/var/log/cinder/api.log)和 cinder-volume(/var/log/cinder/volume.log)。

日志文件 cinder-api 对 endpoint 和连接错误的排查很有帮助。例如,发送请求创建卷设备失败,可以查看 cinder-api 日志文件排查错误。如果该请求已经成功发送并执行或者被禁止,则可以查看 cinder-volume 日志文件排查错误。

**问题 1**:连接调用无效或连接调用在 cinder-api 日志中无记录。

**解决方案**:编辑 nova.conf 文件,添加以下内容。

```
volume_api_class=nova.volume.cinder.API
```

**问题 2**:创建 1GB 的卷设备时,在 cinder-volume 日志文件中报错。

```
2013-03-12 01:35:43 1248 TRACE cinder.openstack.common.rpc.amqp \
ISCSITargetCreateFailed: \
Failed to create iscsi target for volume \
volume-137641b2-af72-4a2f-b243-65fdccd38780.
```

**解决方案**:编辑/etc/tgt/targets.conf 文件,将"include /etc/tgt/conf.d/*.conf"改成"include /etc/tgt/conf.d/cinder_tgt.conf",如下。

```
include /etc/tgt/conf.d/cinder_tgt.conf
include /etc/tgt/conf.d/cinder.conf
default-driver iscsi
```

重启计算节点上的 tgt 和 cinder-*服务,修改生效。

**问题 3**:卷设备映射给虚拟机实例失败,报告找不到 sg_scan 文件。

```
ERROR nova.compute.manager [req-cf2679fd-dd9e-4909-807f-48fe9bda3642 admin
admin|req-cf2679fd-dd9e-4909-807f-48fe9bda3642 admin admin]
  [instance: 7d7c92e0-49fa-4a8e-87c7-73f22a9585d5|instance: 7d7c92e0-49fa-
4a8e-87c7-73f22a9585d5]
  Failed to attach volume  4cc104c4-ac92-4bd6-9b95-c6686746414a at /dev/
vdcTRACE nova.compute.manager
  [instance: 7d7c92e0-49fa-4a8e-87c7-73f22a9585d5|instance: 7d7c92e0-49fa-
```

4a8e-87c7-73f22a9585d5]

    Stdout: '/usr/local/bin/nova-rootwrap: Executable not found: /usr/bin/sg_scan'

**解决方案**：安装 sg3-utils 软件包。

    # zipper install sg3-utils

**问题 4**：在 cinder-volume.log 日志中出现如下错误。

    2013-05-03 15:16:33 INFO [cinder.volume.manager] Updating volume status
    2013-05-03 15:16:33 DEBUG [hp3parclient.http]
    REQ: curl -i https://10.10.22.241:8080/api/v1/cpgs -X GET -H "X-Hp3Par-Wsapi-Sessionkey: 48dc-b69ed2e5
    f259c58e26df9a4c85df110c-8d1e8451" -H "Accept: application/json" -H "User-Agent: python-3parclient"

    2013-05-03 15:16:33 DEBUG [hp3parclient.http] RESP:{'content-length': 311, 'content-type': 'text/plain',
    'status': '400'}

    2013-05-03 15:16:33 DEBUG [hp3parclient.http] RESP BODY:Second simultaneous read on fileno 13 detected.
    Unless you really know what you're doing, make sure that only one greenthread can read any particular socket.
    Consider using a pools.Pool. If you do know what you're doing and want to disable this error,
        call eventlet.debug.hub_multiple_reader_prevention(False)

    2013-05-03 15:16:33 ERROR [cinder.manager] Error during VolumeManager._report_driver_status: Bad request (HTTP 400)
    Traceback (most recent call last):
        File "/usr/lib/python2.7/dist-packages/cinder/manager.py", line 167, in periodic_tasks task(self, context)
        File "/usr/lib/python2.7/dist-packages/cinder/volume/manager.py", line 690, in _report_driver_status volume_stats =
            self.driver.get_volume_stats(refresh=True)
        File "/usr/lib/python2.7/dist-packages/cinder/volume/drivers/san/hp/hp_3par_fc.py", line 77, in get_volume_stats stats =

```
    self.common.get_volume_stats(refresh, self.client)
    File "/usr/lib/python2.7/dist-packages/cinder/volume/drivers/san/hp/hp_
3par_common.py", line 421, in get_volume_stats cpg =
    client.getCPG(self.config.hp3par_cpg)
    File "/usr/lib/python2.7/dist-packages/hp3parclient/client.py", line 231,
in getCPG cpgs = self.getCPGs()
    File  "/usr/lib/python2.7/dist-packages/hp3parclient/client.py",  line
217, in getCPGs response, body = self.http.get('/cpgs')
    File "/usr/lib/python2.7/dist-packages/hp3parclient/http.py", line 255,
in get return self._cs_request(url, 'GET', **kwargs)
    File "/usr/lib/python2.7/dist-packages/hp3parclient/http.py", line 224,
in _cs_request **kwargs)
    File "/usr/lib/python2.7/dist-packages/hp3parclient/http.py", line 198,
in _time_request resp, body = self.request(url, method, **kwargs)
    File "/usr/lib/python2.7/dist-packages/hp3parclient/http.py", line 192,
in request raise exceptions.from_response(resp, body)
    HTTPBadRequest: Bad request (HTTP 400)
```

**解决方案**：更新 hp_3par_fc.py 驱动的副本。

**问题 5**：同一卷设备断开再映射报错。

**解决方案**：使用 nova-attach 命令映射卷设备时，需要修改卷设备名称。当执行完 nova-detach 命令后，卷设备没有从虚拟机实例上删除。此种问题，需要重启 KVM 主机节点解决。

**问题 6**：在计算节点上报如下错误。

```
WARNING nova.virt.libvirt.utils [req-1200f887-c82b-4e7c-a891-fac2e3735dbb\
admin admin|req-1200f887-c82b-4e7c-a891-fac2e3735dbb admin admin] systool\
is not installed
ERROR nova.compute.manager [req-1200f887-c82b-4e7c-a891-fac2e3735dbb admin\
admin|req-1200f887-c82b-4e7c-a891-fac2e3735dbb admin admin]
[instance: df834b5a-8c3f-477a-be9b-47c97626555c|instance: df834b5a-8c3f-47\
7a-be9b-47c97626555c]
Failed to attach volume 13d5c633-903a-4764-a5a0-3336945b1db1 at /dev/vdk.
```

**解决方案**：安装 sysfsutils 软件包。

```
# zipper install sysfsutils
```

**问题 7**：在 FC SAN 环境中，计算节点连接卷设备失败。

```
ERROR nova.compute.manager [req-2ddd5297-e405-44ab-aed3-152cd2cfb8c2 admin\
demo|req-2ddd5297-e405-44ab-aed3-152cd2cfb8c2 admin demo] [instance: 60ebd\
6c7-c1e3-4bf0-8ef0-f07aa4c3d5f3|instance: 60ebd6c7-c1e3-4bf0-8ef0- f07aa4c3\
d5f3]
Failed to connect to volume 6f6a6a9c-dfcf-4c8d-b1a8-4445ff883200 while\
attaching at /dev/vdjTRACE nova.compute.manager [instance: 60ebd6c7- c1e3-4\
bf0-8ef0-f07aa4c3d5f3|instance: 60ebd6c7-c1e3-4bf0-8ef0-f07aa4c3d5f3]
Traceback (most recent call last):…f07aa4c3d5f3\] ClientException: The\
server has either erred or is incapable of performing the requested\
operation.(HTTP 500)(Request-ID: req-71e5132b-21aa-46ee-b3cc-19b5b4ab 2f00)
```

**解决方案**：管理员需要重新检查 WWPN，正确配置 Zone。

**问题 8**：在创建虚拟机实例时，实例的状态由 BUILD 变成 ERROR。

**解决方案**：在计算节点上查看 cat/proc/cpuinfo 配置文件是否设置 VMX 或 SVM。同时，确认 CPU 是否支持虚拟化。